Discovering Dynamical Systems Through Experiment and Inquiry

Textbooks in Mathematics

Series editors:

Al Boggess, Kenneth H. Rosen

Advanced Calculus
Theory and Practice, Second Edition
John Srdjan Petrovic

Advanced Problem Solving Using Maple
Applied Mathematics, Operations Research, Business Analytics, and Decision Analysis
William P Fox, William Bauldry

Nonlinear Optimization
Models and Applications
William P. Fox

Linear Algebra
James R. Kirkwood, Bessie H. Kirkwood

Train Your Brain
Challenging Yet Elementary Mathematics
Bogumil Kaminski, Pawel Pralat

Real Analysis
With Proof Strategies
Daniel W. Cunningham

Contemporary Abstract Algebra, Tenth Edition
Joseph A. Gallian

Geometry and Its Applications, Third Edition
Walter J. Meyer

Linear Algebra
What You Need to Know
Hugo J. Woerdman

Introduction to Real Analysis, Third Edition
Manfred Stoll

Discovering Dynamical Systems Through Experiment and Inquiry
Thomas LoFaro and Jeff Ford

For more information about this series, please visit: https://www.routledge.com/
Textbooks-in-Mathematics/book-series/CANDHTEXBOOMTH

Discovering Dynamical Systems Through Experiment and Inquiry

Thomas LoFaro and Jeff Ford

CRC Press
Taylor & Francis Group
Boca Raton London New York

CRC Press is an imprint of the
Taylor & Francis Group, an **informa** business

First edition published 2021
by CRC Press
6000 Broken Sound Parkway NW, Suite 300, Boca Raton, FL 33487-2742

and by CRC Press
2 Park Square, Milton Park, Abingdon, Oxon, OX14 4RN

© 2021 Taylor & Francis Group, LLC

CRC Press is an imprint of Taylor & Francis Group, LLC

ISBN: 978-0-367-90394-7 (hbk)
ISBN: 978-0-367-71376-8 (pbk)
ISBN: 978-1-003-02413-2 (ebk)

Typeset in Computer Modern font
by KnowledgeWorks Global Ltd

In memory of
J.R. and T.J.

Contents

Preface

This textbook was written primarily for undergraduate mathematics students who have taken the entirety of calculus and have completed an introduction to proofs course. Ideally, students would also have taken linear algebra and an introductory analysis course, but this is not necessary. For the most part, the required material from these disciplines is included in the text.

This book differs from most texts on dynamical systems by blending the use computer simulations with inquiry-based learning (IBL). Computer simulations have long been used in the study of dynamical systems to visualize chaotic behavior, and we believe that these tools provide students with a platform from which they can begin to see more general mathematical truths.

Inquiry-based learning is an excellent tool to move students from merely remembering the material, to deeply understanding and analyzing the mathematics. The method teaches students to think, to question, and to create, rather than memorize. The method relies on asking students questions first, rather than presenting the material in a lecture. Students are expected to work out problems on their own or in groups, and then the material is discussed in class. The instructor is less of a "sage-on-the-stage," and more of a "mentor-in-the-middle," giving students the room to explore the content, while gently guiding them toward the correct solutions.

Inquiry-based learning has been shown to improve student learning and to be a more equitable method of content delivery. Students in an IBL classroom tend to have longer retention of the material [5]. Even the students, who may have trouble recalling specifics about the material have learned how to learn, and that skill carries over to all disciplines of study. We also see that underrepresented groups perform better in the IBL classroom [6]. IBL gives students a chance to proceed at their own pace, to make mistakes, to learn from them, and to succeed on their own terms.

There are two main ideas that we find helpful as IBL instructors. The first is to be less helpful. It can be tough to watch a student struggle at the board, but our jobs as instructors is to make that struggle productive. Ask questions to lead them to the right solutions, without giving away the solutions. We have attempted to scaffold the activities in this book so that students need only make small leaps in their understanding as they progress. Small hints are occasionally provided when the proof requires something unusual or uses ideas they learned in other courses.

The second idea is that of productive failure. Students will have many false starts and wrong turns in their understanding of the content. It is our job to ensure that they see those failures as learning opportunities, rather than disappointments. Be patient, be kind, but hold the students accountable. You can expect some resistance to the method for a few weeks. We have had students complain: "He doesn't teach anything; he just stands there asking questions." This is normal, but once a student experiences the joy of self-discovery, there is no turning back.

If you are new to using IBL methods in the classroom, some helpful resources are available from the Academy of Inquiry Based Learning (http://www.inquirybasedlearning.org/) and the Legacy of R.L. Moore (http://legacyrlmoore.org/). Both of these organizations offer workshops to help get instructors involved with IBL. The Academy of Inquiry Based Learning, in particular has online resources and many active communities for instructors new to the method.

We refer to the unique approach to teaching mathematics used in this book as ECAP for **E**xplore, **C**onjecture, **A**pply, and **P**rove. We believe that ECAP provides students with a deeper understanding of dynamical systems because it mimics the actual practice of mathematics. In general, each section begins with exercises guiding students through explorations of the featured concept in dynamical systems and concludes with exercises that help the student formally prove the result.

We chose to put the explorations first (as opposed to end-of-chapter projects) because in our experience, computer explorations provide students with an enhanced understanding of the mathematical concepts by allowing improved visualization that is both dynamic and engaging. Moreover, the ability to manipulate the equations and/or underlying mathematical aspects allows students to grasp what aspects are essential for a given phenomenon and what are not. This exploration, combined with some guided questions, gives students the confidence to conjecture the formal theorems being formed. The proof subsection then completes the student's journey from the discovery of the mathematics to the final step of proving the validity of the result.

Let's look at this structure in more detail to understand how this text might be used during the semester. Most sections of the text are divided into **Explore, Conjecture, Apply,** and **Prove** portions. We find this to be an effective approach to guide the young mathematician toward the depth of understanding expected in an IBL classroom.

The **Explore** sections ask leading questions to introduce a concept and help students see what may be true. Many of the explorations use numerical activities that we have designed specifically to highlight the concepts and ideas being explored. Each accompanying activity is available on `IBLdynamics.com`. There you will find a collection of interactive apps designed to get students thinking about the topics at hand. Students can discover examples and counterexamples through manipulations built into the software. This is a unique feature of the text, and one that has proven essential for students taking the course in the past.

We recommend assigning as many of the **Explore** problems as you think are feasible for your course. These will mostly be done outside of class, but it is worthwhile to do some of these activities in pairs or small groups during class-time. This is especially true early in the term when students are unfamiliar with inquiry-based learning. There is value in having multiple students share different approaches to solving a problem, especially when some are incorrect.

Following the explorations, we move on to the **Conjecture** sections. These activities are most effective when used in the classroom as prompts for discussion about the mathematics. Encourage multiple students to share their ideas. Look for commonalities and disagreement in their responses. Prompt the students to share how the explorations led them to believe that certain things are true. Look for students who have reached different conclusions. Keep asking questions until students question the validity of a false conjecture, or are more deeply convinced of a true conjecture.

The **Apply** portions include many problems to help students confirm or disprove their conjectures and to give them the opportunity to practice the technique at hand. Assign what you feel is reasonable to your students. It is important to remember that students may also be doing **Exploration** or **Proof** exercises and that all of this work needs to be carefully balanced.

Once the students have explored the content, conjectured the generalities, and applied their results, they should move on to the proofs in the **Proof** portion of the section. We have had success assigning some of these proofs to be presented in class, and some to be turned in as homework. You will need to balance your students' needs with the relevance of the mathematics to determine what proofs should be assigned as homework vs. presented and discussed in class. When the proofs are presented in class, students should be encouraged to present their proofs. These should be again used to foster mathematical discussions. It is also worthwhile to have a student "recorder," who transcribes this work and writes a final version of the proof to be shared with the class. As you move into more difficult proofs, more scaffolding is provided to help guide the students.

Advice for Using this Text

When determining what material to include in this text, we focused primarily on how that topic would or would not lend itself to the ECAP structure. For example, there is no discussion of complex dynamics because we felt that going from explorations of Julia sets to conjectures and proofs about their connectivity would be unreasonable for most undergraduate students. There are certain topics that are included that we felt were essential to explore and conjecture, but chose not to prove either because of its difficulty or because a complete proof would require deeper ideas from other areas of mathematics. A prime example of this is the period-doubling bifurcation theorem.

On the other hand, we did see a topical theme emerge that we believe is unique to a dynamical systems text: symbolic dynamics. While symbolic dynamics is a fairly standard topic in an undergraduate dynamics text, we have tried to emphasize it in a way that is more detailed and inclusive than is typically the case. This is partly because the ideas and proofs in this area are accessible to undergraduates and approachable using the ECAP methodology. Thus we chose to conclude this book with a chapter dedicated to a general discussion of the topic.

Finally, we have chosen to include multiple sections on important ideas from analysis and topology independent from their application to dynamics. We feel that most students need a refresher (at the least) on these topics and the study of dynamical systems provides a wonderful motivation and/or introduction to these ideas.

In piloting this material, we were able to cover the first 10 chapters of this book in a single semester. If you don't need to cover the analysis material, you will most likely be able to cover at least some of the material in Chapters 11–13. We believe that the pairings of Chapters 11 and 12 or of Chapters 12 and 13 work well. Chapter 11 is not a prerequisite for Chapters 12 and 13 though the material concerning the Hénon map in Chapter 11 nicely illustrates the theme of Chapter 12.

Chapter 1 defines discrete dynamical systems and discusses numerical methods for tracking the orbits of points in a system. It also introduces the students to some of the tools on $\boxed{\texttt{IBLdynamics.com}}$. Students can also work on graphical iteration, and see how dynamics can apply to population models.

In our experience, the upper-level students taking this class have little experience with the analysis of sequences and other tools from analysis that arise when studying dynamical systems. Thus, we have chosen to include much of this material in this book. Chapter 2 provides a fairly complete coverage of sequences for this reason. You may choose to skip this chapter if that is not the case with your students.

Chapter 3 defines fixed and periodic points. The number of explorations on iteration increases here, as we work to build a students' intuition about attracting and repelling points. They are challenged to make conjectures about attracting and repelling fixed points, studying linear, quadratic, and logistic maps. Population models are revisited and some important theorems about fixed and periodic points are proven.

Much of Chapter 4 covers analysis topics with the primary focus being the implicit function theorem and fixed point theorems. The students are given more opportunity to come up with counterexamples, and the explorations reflect this. The chapter concludes with a proof of the hyperbolic fixed points theorem. Section 4.2 can be safely omitted if your students are familiar with the implicit function theorem.

Bifurcations are covered in Chapter 5. A large number of explorations are included at the beginning of the chapter to drive students toward conjecturing important theorems about bifurcations. Tangent and period-doubling bifurcations are covered, as well as bifurcation diagrams.

Chapters 6 and 7 introduce the difference between local and global dynamics. Students explore tools to catalog all of the different behaviors that can occur in a system. The logistic map and the doubling map are studied. Symbolic dynamics is introduced in the form of binary sequences to study orbits in the doubling map. The logistic map is revisited with coefficients greater than or equal to 4.

In Chapter 7, we delve further into symbolic representations, particularly the shift space on two symbols. The Cantor Set is constructed to aid with motivation, and the definition of a metric is introduced. If Chapter 2 was omitted, portions may need to be included here, depending on student background.

Chaos is defined in Chapter 8. Interactive portions of the website help illustrate sensitive dependence on initial conditions and topological transitivity. The shift map, the doubling map, and the logistic map are returned to as examples.

Chapter 9 presents bifurcation diagrams for the logistic map to demonstrate the transition to chaotic behavior. Interactive tools help students find windows of stable periodic behavior, and a study of the ordering of periodic windows is introduced. This lays the foundation for Chapter 10.

Sarkovskii's theorem and its proof is the focus of Chapter 10. There are many explorations and interactive activities to motivate the theorem. The chapter begins with a short review of the intermediate value theorem and two fixed point theorems from Chapter 4. Then Sarkovskii's theorem is discussed. The student is guided through a partial proof at the end of the chapter.

Chapter 11 introduces dynamical systems in the plane. Section 11.1 may be omitted if the class has sufficient background in linear algebra. The chapter starts with linear systems, moves to systems with complex eigenvalues, and then on to nonlinear systems. Interactive explorations on the website illustrate the gingerbread attractor and the Hénon map. Instructors may wish to review multivariable calculus or basic complex arithmetic as necessary.

The Smale horseshoe is motivated and constructed in Chapter 12. As a tool to study the horseshoe, two-sided symbolic spaces and their associated shift maps are introduced. The chapter concludes by relating the horseshoe to the Hénon map.

The final Chapter, 13, goes into more detail about symbolic dynamical systems, and the tools to study them. Symbolic systems on n-symbols are defined, as are shifts of finite type. We introduce graph theory as a way to study conjugate systems. Applications are given to tiling spaces and Markov chains. The chapter concludes with an exploration of Markov partitions and Arnold's cat map.

Message to Students

We could have structured the content of this text like a typical mathematics book, but we've chosen not to, and for a specific reason. This book is designed to help you learn to think like a mathematician. We want you to go beyond memorizing theorems and calculating results. We want you to explore, to discover, and to prove the results for yourself. Most students are brought up with the idea that mathematics is entirely about showing what is true. It's actually the case that the much more important step is showing when and why something is true. In this text, we plan for you to achieve exactly that.

Most every section starts with exploration activities. You will be asked to work things out on paper or use activities at IBLdynamics.com to explore some topic in dynamical systems. It is expected that you try everything. Take notes, notice patterns, and don't be afraid to make mistakes. If you see something that seems to be generally true, be ready to make that conjecture. Be prepared to share your insight with the class. And be ready to explain why you think your insight is true. Your instructor will guide you through the process, but ultimately, the work must be your own. You will learn that it is acceptable to fail, and that making a false conjecture, or an error in explanation, is just part of learning to be a mathematician. You will learn that these failures can be productive. If we learn from our errors and move forward, an error is not a failure, it is an opportunity to truly understand something new.

A course like this returns dividends proportional to your efforts. It is much easier to be given an algorithm, and to execute that algorithm, than to struggle for the results yourself. It is difficult, but struggling is essential for understanding the mathematics and for personal growth. We sometimes tell our students to "embrace the stuck" because all new and important ideas originate here.

The joy of discovery, the frustration of exploration, and the satisfaction of collaboration are all important aspects of mathematics. You will explore. You will conjecture. You will prove results. And you will apply results to a variety of related problems. All of these aspects combine when one is creating new mathematics. It is the intention of the authors that this course will prepare you, not just for an understanding of dynamical systems, but for exploration into any area of mathematics you find appealing.

Acknowledgments

We have received the help and support of many. But foremost on that list are our colleagues in the MCS Department at Gustavus Adolphus College who have listened to our musings, critiqued our writing, and occasionally covered our classes. We would be remiss if we didn't single out Mike Hvidsten, who encouraged us to take this project from a collection of classroom activities and demonstrations to a full-fledged textbook.

We had the help of two wonderful Gustavus students, Emilee Mason and Zach Dawson. Emilee, who "beta-tested" this project in our Discrete Dynamical Systems course, was invaluable in her proofreading and editing duties. She was especially helpful in finding all of the places where we omitted commas! If any are still missing, it is our fault and not hers. Zach Dawson was a lifesaver! A computer science major, Zach translated all of Tom's poor *Mathematica* code into clean and functioning *JavaScript*. We would still be struggling with the basics of *JavaScript* if it weren't for Zach.

Tom would also like to thank Francis Su and his colleagues at Harvey Mudd College. During the infancy of this project, Tom spent a week visiting Francis picking his brain about writing, inquiry-based learning, and life. As is often the case, these latter conversations were the most valuable.

Jeff would like to thank Dr. Steven Clontz at University of Southern Alabama, and the late Dr. Frank Sturm, for their endless late-night discussions about the value of inquiry-based learning and the nuts and bolts of how to make it effective.

Various aspects of this work, including funding for Emilee and Zach, were possible through a generous gift to Gustavus Adolphus College by the family of Clifford M. Swanson.

1

An Introduction to Dynamical Systems

1.1 What Is a Dynamical System

At the most basic level, a dynamical system is simply something that evolves deterministically over time. In this context, deterministic means that the system changes by rules that are fixed and not random. Dynamical systems arise in an incredibly wide variety of applications. For example, the motion of a pendulum is a continuous dynamical system with the angular position and angular momentum determined at every time. If we know the initial position and momentum, then the rules of the dynamical system determine these quantities at every time in the future. In ecology, models of population growth are often discrete dynamical systems. The model is a function that uses the population at one generation to compute the population of the next generation. Once again, if we know the initial population then by repeatedly applying this rule we can compute the population at any time in the future.

Ordinary differential equations are often the vehicle for modeling *continuous dynamical systems* like the pendulum model. In this book we focus on *discrete dynamical systems*.

Definition 1.1 *A* **discrete dynamical system** *is a function* $f : X \to X$ *that generates a sequence*

$$\{x_n\}_{n=0}^{\infty} = \{x_0, x_1, \dots\}$$

from an initial condition $x_0 \in X$ *using the rule* $x_{n+1} = f(x_n)$.

The space X is called the *state space* of the dynamical system, and for most of this book, X will be the real numbers $\mathbb{R}$. What is important here is that it is very easy to compute the sequences generated by discrete dynamical systems. Starting with an initial condition x_0, we just apply the function f to compute x_1. In other words,

$$f(x_0) = x_1.$$

The subscript can be thought of as a discrete time measurement. If time is measured in years, then x_0 is the value of x in year 0, x_1 is the value of x in year 1, and so on.

Once we have computed x_1 we then compute x_2 by applying f to x_1 so that

$$f(x_1) = x_2.$$

By continuing this process indefinitely, we generate the sequence $\{x_n\}_{n=0}^{\infty}$. This process of starting with an initial condition x_0 and repeatedly applying the function f is called **iteration.**

One would think that such a simple system would only exhibit relatively straightforward behaviors. Maybe solutions can only approach an equilibrium or at worst be periodic. This is far from true. The kinds of behaviors that can occur are quite remarkable; so remarkable that we often refer to these dynamics as *chaotic dynamics.* Even more remarkably, this wild and chaotic behavior, which at first seems almost beyond description, can often be easily described using just sequences of zeros and ones!

The purpose of this text is to guide you on a journey where you discover these behaviors, you conjecture mathematical statements about them, and ultimately you prove the truth of these mathematical statements. Let's pack our gear and begin exploring.

1.2   Numerical Iteration and Orbits

Iteration is best illustrated with a simple example. Let's consider the function $f(x) = x/2$ and take as an initial condition $x_0 = 12$. Then

$$x_1 = f(x_0) = f(12) = 6$$

and

$$x_2 = f(x_1) = f(6) = 3$$

and so on. Tables, such as the one inset into figure 1.1, are a natural way to show sequences generated by iteration. The first column is the iteration number, and the second column is the sequence of iterates of f with $x_0 = 12$.

The plot in figure 1.1 shows the same iterations but in a graphical format. Here the horizontal axis is the iteration number (n), and the vertical axis, the sequence element (x_n).

Notice that each entry in the second column of the table in figure 1.1 is f applied to the entry immediately above it. This is iteration and the focus of our study. Given a dynamical system, we would like to describe all of the possible sequences that are generated by the given function. We call these sequences **orbits.** The example in figure 1.1 shows the first six terms of the orbit of $x_0 = 12$ under iteration by $f(x)$. We might describe this particular orbit by saying

the orbit of $x_0 = 12$ under iteration by $f(x) = x/2$ converges to $x = 0$.

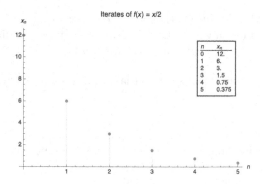

FIGURE 1.1
The first six iterates of $f(x) = x/2$ with an initial condition of $x_0 = 12$
displayed as both a table and a graph.

You probably already see that every orbit of this dynamical system, regardless of initial condition, converges to 0 and by doing so you've completely
described this dynamical system.

Explore

Exploration 1. Consider the dynamical system

$$x_{n+1} = x_n^2.$$

For each of the initial conditions listed below, compute the first 10 iterates of
their orbit. Make both a table and a plot like those in figure 1.1. In each case,
describe the orbit in your own words. Look for commonalities and differences
in the different orbits and try to categorize the different types. **Suggestion:**
Doing this by hand gets old pretty quick. However, doing this with a spreadsheet is very easy. After doing the first couple by hand, think about how you
might implement this dynamical system in a spreadsheet program. This can
also be used to create plots such as that show in figure 1.1.

a)	$x_0 = -2.0$		b)	$x_0 = -1.5$		c)	$x_0 = -1.0$
d)	$x_0 = -0.5$		e)	$x_0 = 0.0$		f)	$x_0 = 0.5$
g)	$x_0 = 1.0$		h)	$x_0 = 1.5$		i)	$x_0 = 2.0$

Exploration 2. Repeat exploration 1 using the dynamical system

$$x_{n+1} = x_n^2 - 1.$$

Use the same initial conditions that you used in exploration 1. Describe each of the orbits in your own words. Look for commonalities and differences in the different orbits and try to categorize the different types.

Exploration 3. Using the same initial conditions as exercise 1, find the orbits of

$$x_{n+1} = x_n^2 + 1.$$

Describe the orbits in your own words. Look for commonalities and differences in the different orbits and try to categorize the different types.

Exploration 4. Explore the dynamical system

$$x_{n+1} = 1 - |x_n|.$$

Use the initial conditions on the following table and describe the orbits.

a)	$x_0 = -1.0$	b)	$x_0 = -0.5$	c)	$x_0 = -0.4$
d)	$x_0 = -0.2$	e)	$x_0 = 0.0$	f)	$x_0 = 0.2$
g)	$x_0 = 0.4$	h)	$x_0 = 0.5$	i)	$x_0 = 1$

Exploration 5. For each of explorations 1 through 3, find an initial condition where the orbit only contains one value, or show that no such point exists.

Before going much further, we need to introduce some basic terminology that we will be using to describe orbits. The first and most important type of orbit is one that is constant or fixed. For example, the orbit of $x_0 = 0$ is fixed in exploration 1.

Definition 1.2 *Let $f : X \to X$. A point $\tilde{x}$ is a* **fixed point** *of f if*

$$f(\tilde{x}) = \tilde{x}. \tag{1.1}$$

Fixed points are also referred to as *equilibrium points*. It follows directly from this definition that if $\tilde{x}$ is a fixed point, then the orbit of $\tilde{x}$ is the sequence $\{\tilde{x}, \tilde{x}, \tilde{x}, \dots\}$ since each application of the function f leaves $\tilde{x}$ unchanged.

One almost always begins the study of a given dynamical system by computing the fixed points. This is such an important concept that a good portion of chapter 3 is dedicated to the subject.

In exploration 2 you saw some other interesting behavior when the initial condition $x_0 = 0$. You observed that

$$f(0) = -1, \ f(-1) = 0, \ f(0) = -1, \dots.$$

We can restate this another way. The orbit of $x_0 = 0$ under iteration by f is

$$\{0, -1, 0, -1, \dots\}.$$

We call this a period 2 orbit or we might say that $x_0 = 0$ is a **period 2 point** of this dynamical system. Of course $x_0 = -1$ is also a period 2 point.

Definition 1.3 *Let* $f : X \to X$. *A point* x_0 *is a* **period 2 point** *of the dynamical system* $x_{n+1} = f(x_n)$ *if*

$$f(f(x_0)) = x_0.$$

If $f(x_0) \neq x_0$, *then we say that* x_0 *is a* **prime period 2 point.**

Let's discuss this definition a bit further. First, note the two-fold composition in the definition. Iteration is ultimately the repeated composition of a function with itself since the output of one step becomes the input to the next. If $f(x_0) = x_1$ and $f(x_1) = x_0$, then

$$f\left(f\left(x_0\right)\right) = f(x_1) = x_0.$$

In other words, and this is important, a period 2 point of $f(x)$ is a fixed point of $f(f(x))$. Also note that a fixed point $\tilde{x}$ of f must also be a period 2 point of f since in this case

$$f\left(f\left(\tilde{x}\right)\right) = f(\tilde{x}) = \tilde{x}.$$

To distinguish fixed points from "true" period 2 points, we use the adjective "prime."

Dynamical systems can also have period 3 points, period 4 points, and so on. A point x_0 is a period 3 point if

$$f\left(f\left(f\left(x_0\right)\right)\right) = x_0.$$

Obviously this gets a bit cumbersome notationally. Imagine what you would have to write down to describe a period 243 point! To deal with this problem, we use superscripts on the function name. More formally, $f^n(x)$ denotes the n-fold composition of f with itself. So

$$f^3(x) = f(f(f(x)))$$

and so on.

A Warning about Composition Notation

It is natural to read $f^3(x)$ as "$f(x)$ cubed" or "$f(x)$ to the third power." **This is wrong!** Always remember that the superscript refers to function composition and not exponentiation.

Exploration 6. What is $f^n(f^m(x))$?

Using this notation, we can now define periodic points in general.

Definition 1.4 *Let* $f : X \to X$. *A point* x_0 *is a* **period** p **point** *of the dynamical system* $x_{n+1} = f(x_n)$ *if*

$$f^p(x_0) = x_0.$$

If $f^k(x_0) \neq x_0$ *for all* $k < p$, *then we say that* x_0 *is a* **prime period** p **point.**

Exploration 7. IBLdynamics.com The tool for this exploration on the website gives the first 11 elements of orbits to several dynamical systems. Determine whether they are periodic or not. If so, determine the period.

1.3   Graphical Iteration

There is a very useful way of thinking about the iteration of the discrete dynamical system $x_{n+1} = f(x_n)$ that uses the graph of the function $y = f(x)$ and the diagonal line $y = x$. Recall that the basic idea of iteration is that the output from step n becomes the input to step $n+1$. If we have a plot showing the graph of $y = f(x)$ and the line $y = x$, then we can find $x_1 = f(x_0)$ by drawing a *vertical line* from x_0 on the x-axis to the graph of $y = f(x)$ to find the point $(x_0, f(x_0)) = (x_0, x_1)$. Note that the output x_1 is the y-coordinate of this pair. We next need to convert the output x_1 from a y-coordinate to an x-coordinate so that we can input it to f to find $x_2 = f(x_1)$. Thus the next step in this process is to locate x_1 on the x-axis. We do this by drawing a *horizontal line* from the point (x_0, x_1) to the line $y = x$. The coordinate of this point is (x_1, x_1) since all points on a horizontal line all have the same y-coordinate and we stopped drawing where $y = x$. Now going *vertically* to the x-axis gives us the location of the value x_1 on the x-axis. Finally, going *vertically* again to the graph of $y = f(x)$ gives us x_2 as a y-coordinate. We repeat this process to iterate the dynamical system and in so doing graphically compute the orbit of x_0.

Note that at one point in the process, we go vertically to the x-axis and then vertically to the graph of $y = f(x)$. This seems wasteful and is in fact unnecessary. Here is a step-by-step description of this process that eliminates this wasteful step.

1. Draw a vertical line to the graph of $y = f(x)$.

2. Draw a horizontal line to the diagonal $y = x$.

3. Repeat, repeat, repeat, ...

This process gives a picture like the one in figure 1.2 that shows the iteration of $x_{n+1} = x_n^2$ with an initial condition of $x_0 = 0.9$. Note how the "stairsteps" start at the point $(0.9, 0.81)$ on the graph of $y = x^2$ and go *horizontally* to the diagonal $y = x$ and then *vertically* to $y = x^2$. Each vertical line represents one iteration. Hence this figure shows a total of 6 iterates (the first vertical line from the x-axis to the graph is not shown which is why you only count 5 of them).

Graphical iteration is sometimes referred to as **graphical analysis** and the pictures generated in this way are called either **stair-step diagrams** or **cobweb diagrams**.

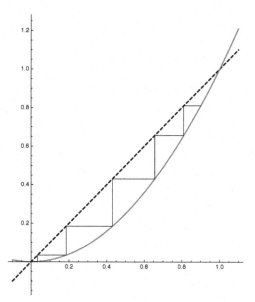

FIGURE 1.2
A plot showing the graphical analysis of $x_{n+1} = x_n^2$ with an initial condition of $x_0 = 0.9$.

Explore

Exploration 8. Explain why a fixed point of a function f is the intersection of the graph of $y = f(x)$ and the line $y = x$.

Exploration 9. ⎢IBLdynamics.com⎥ Print out the graphs in figure 1.3 from the IBLdynamics.com website. For each graph, approximate the values of the fixed points (if any). Practice generating stair-step diagrams for each of these dynamical systems. For each one, describe in your own words the orbits that you find. Look for commonalities and differences in the different orbits and try to categorize the different types.

Exploration 10. ⎢IBLdynamics.com⎥ Adjust the graph at IBLdynamics.com for each of the following parameters (i.e., k values). Draw the graph on paper and construct a stair-step diagram with the indicated initial conditions.

k	0.5	0.25	0.1	0.1	−0.5
x_0	0.5	0.5	0.4	0.6	0.5

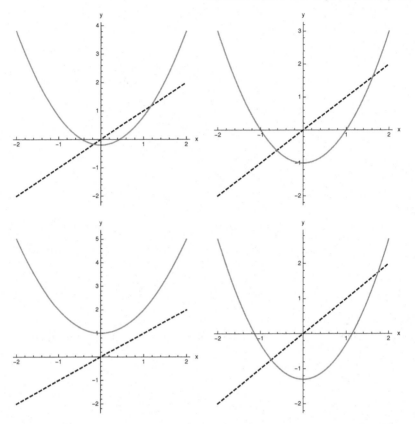

FIGURE 1.3
Figures for practicing graphical analysis.

1.4   Modeling Using Discrete Dynamical Systems

Discrete dynamical systems are incredibly useful for modeling real-world phenomena that occur in discrete time steps. In this section you will develop a population dynamics model that falls into this category. Let p_n denote the population of some species at generation n.

Apply

Assumption: Generations are non-overlapping, and the population at generation $n+1$ depends deterministically on only the population at generation n. In other words, there exists some function f such that $p_{n+1} = f(p_n)$.

Framework: Assume that the function f has the form

$$f(p) = \lambda(p)V(p)p$$

where

- $V(p)$ is called the **viability** and measures the percentage of newborns in the species that survive to reproductive age, and

- $\lambda(p)$ is the **mean reproductive rate** and measures the mean number of newborns per adult.

Application 11. Given the above descriptions of the functions λ and V, what mathematical properties do you think each of these functions should have? You might want to first think about what values they should take. Then about what shapes the graphs should or should not have. For each property you list, explain how and why that property is related to the modeling of a population. This might mean listing additional modeling assumptions.

Application 12. Give one or two examples of possible viability functions $V(p)$. This function may also use some parameters (i.e., constants), but you do not need to assign specific values to these parameters. Are there any underlying assumptions inherent in your choice of V that you did not specify in application 11? **Suggestion:** Start with the simplest example you can think of and then build from that.

Application 13. Give one or two examples of a possible reproductive rate function $\lambda(p)$. This function may also use some parameters (i.e., constants), but you do not need to assign specific values to these parameters. Are there any underlying assumptions inherent in your choice of λ that you did not specify in application 11? **Suggestion:** Start with the simplest example you can think of and then build from that.

Application 14. Use the methods described in this chapter to play with your model. Try a variety of reasonable parameter values for the unassigned parameters. We suggest using a spreadsheet to do this, but feel free to use whatever tool you find appropriate. Describe some of the behaviors that your model produces. Below are a few questions that might be helpful to you as you do this. Do not limit yourself to these questions. Be creative!

1. Is the behavior that you observe realistic? For example, does it ever lead to negative population numbers. If it is not realistic, think about modifying it and repeating this process.

2. Does the population approach some fixed value? Does it oscillate between multiple values? Does it do something more complicated?

3. Does the behavior depend on the initial population size in some way? For example, does the population go extinct for some starting values but not for others?

4. How does the behavior change, if at all, when you change some of the parameters in the model?

5. Graph your function $f(p) = \lambda(p)V(p)p$ and the diagonal line as you did in section 1.3. Describe how this graph changes as you adjust one or more of the parameters in the model.

2

Sequences

2.1 Introduction to Sequences

The study of discrete dynamical systems is ultimately the study of infinite sequences generated via iteration as described in chapter 1. You have probably spent some time in your mathematics career learning about sequences. For example, in calculus 2, you spent several days learning about infinite sequences and how they are used to study infinite series. Or maybe you played with the Fibonacci sequence as a child. And if you had an analysis class, maybe you learned how sequences of rational numbers are used to define the real numbers. In this class, we are studying sequences not as a vehicle to understand some other mathematical concept, but rather as objects that are interesting in their own right.

This chapter provides some of the fundamental concepts for studying sequences and while it may be review for many of you, we believe that spending some time on the fundamentals is necessary before returning to discrete dynamical systems.

Explore

Let's begin with a fun example, the Fibonacci sequence. The first two terms of the Fibonacci sequence are $F_0 = 1$ and $F_1 = 1$. Each subsequent term is determined by the recurrence

$$F_n = F_{n-1} + F_{n-2}$$

so that each term is simply the sum of the previous two terms.

Exploration 1. Compute F_n for $n = 0, \ldots, 10$.

It is natural (and useful) to think of a sequence as simply an ordered list of things (usually numbers). As mathematicians, we need to turn this intuitive idea into something much more precise.

Definition 2.1 *A sequence* $\{x_0, x_1, x_2, \dots\} = \{x_n\}_{n=0}^{\infty}$ *with* $x_n \in X$ *for all* n *is a function from the natural numbers* $\mathbb{N}$ *to the space* X.

It is common practice to omit the $n = 1$ and the ∞ from this notation and simply write $\{x_n\}$ when referring to a sequence when no confusion will arise.

In this more sophisticated view, a sequence is a function that assigns a value in the space X to each natural number via $n \mapsto x_n$. This gives a list of numbers with the position of the number in the list corresponding to n. The space X is the set of all outputs of this function (or codomain) and for us X will almost always be the real numbers $\mathbb{R}$.

Exploration 2. Compute the first 5 elements of each of the sequences below. Describe the behavior of each sequence in your own words.

1. $x_n = 1 - \dfrac{1}{n}$

2. $y_n = (-1)^n$

3. $z_n = \left(\dfrac{4}{5}\right)^n$

4. $w_0 = 0,\ w_{n+1} = \dfrac{4}{5}w_n$

5. $a_n = \dfrac{n^2 + n - 1}{n + 4}$

6. $b_n = \dfrac{n^2 + n - 1}{n^2 + 4}$

2.2   Convergence of Sequences

The most basic question you can ask about a sequence is whether it converges or not. You have an intuitive understanding of what this means. A sequence converges to some value L if the terms of the sequence get "closer and closer to L."

But because logical precision is essential in mathematics, this intuitive idea needs to be formalized. Here is the formal definition where we assume that $x_n \in \mathbb{R}$ for all n.

Definition 2.2 *A sequence* $\{x_n\}$ **converges** *to a limit* $L < \infty$ *if for every* $\varepsilon > 0$ *there exists an* $N > 0$ *such that* $n > N$ *implies that* $|x_n - L| < \varepsilon$. *When the sequence* $\{x_n\}$ *converges to a limit* L *we write*

$$\lim_{n \to \infty} x_n = L.$$

If a sequence does not converge to a limit L, *we say that the sequence* **diverges**.

This definition should sound very familiar to you. It is almost exactly the definition of a limit that you learned in calculus. In particular, it is almost word for word the definition of a "limit at infinity." How does definition 2.2 formalize our intuitive concept of convergence? If ε is how closely you want to approximate the limit L, then N is how far you have to go out in the sequence to obtain this level of precision.

Understanding Absolute Values

One of the main difficulties for many students when using the definition of a limit is dealing with absolute values. Many of us were told that when seeing an absolute value we simply "drop the minus sign." Unfortunately, this simple rule is not at all helpful when dealing with expressions like the one in definition 2.2.

Here is a more helpful way of interpreting the expression

$$|A - B|. \tag{2.1}$$

Read expression 2.1 as "the distance between A and B." The inequality

$$|x - 4| < 3$$

then reads "the distance between x and 4 is less than 3." When thought of this way, it is clear that this is the same as

$$1 < x < 7$$

where the left-hand term is 1 because $4 - 3 = 1$ and the right-hand term is 7 because $4 + 3 = 7$.

Using this idea, the inequality

$$|x_n - L| < \varepsilon$$

in definition 2.2 reads "the distance between x_n and L is less than ε." This naturally leads to

$$L - \varepsilon < x_n < L + \varepsilon$$

and we now have the expression written in a form that is amenable to algebra.

Explore

Exploration 3. Interpret each of the following absolute value statements using the distance interpretation described in the box above. Express each of them as a double inequality with only x in the center. You may need to manipulate some of them to put it in the necessary form.

a)	$\|x - 3\| < 0.1$	b)	$\|x + 3\| < 0.1$	c)	$\|x\| < 0.1$
d)	$\|3x - 6\| < 1$	e)	$\|10x - 5\| < 2$	f)	$\|6x - 4\| < 10$

Exploration 4. $\boxed{\texttt{IBLdynamics.com}}$ The tool on the website for this exercise shows the elements of the sequence

$$x_n = 1 - \frac{1}{n},$$

the limit value is $L = 1$, and the "ε-band" is the interval $(L - \varepsilon, L + \varepsilon)$ about $L = 1$. The tool allows you to change the value of ε. Doing this causes the image to zoom in or out vertically (by changing the limits on the vertical axis) and to zoom in or out horizontally (by changing the limits on the horizontal axis). Use the tool to estimate the value of N needed to ensure that all further elements of the sequence are in the ε-band for 5 different values of ε. Do you see a relationship between your chosen value of ε and N?

Exploration 5. Reread definition 2.2. For the values of ε that you chose in exploration 4, directly compute the value of N so that

$$|x_n - L| < \varepsilon.$$

How do each of these computed values compare with your answers in Exploration 4?

Apply

It is often the case that the techniques you used in calculus to compute limits at infinity apply to sequences. That is the case with many of the exercises in this section. It might be helpful to review this material in general, and L'Hopital's rule in particular.

Application 6. Compute the limit of the sequence $z_n = \left(\dfrac{4}{5}\right)^n$ and use definition 2.2 to prove that it converges to this limit value.

Application 7. Compute the limit of the sequence $z_n = a^n$ for $0 < a < 1$ and use definition 2.2 to prove that it converges to this limit value.

Application 8. Compute the limit of the sequence $z_n = \dfrac{2n + 4}{5n + 1}$ and use definition 2.2 to prove that it converges to this limit value.

Application 9. Compute the limit of the sequence $z_n = \dfrac{2}{5n + 1}$ and use definition 2.2 to prove that it converges to this limit value.

Application 10. Compute the first 10 terms of the sequence

$$z_n = (-1)^n + \frac{1}{n}.$$

Does the sequence converge or diverge? Why?

Application 11. Compute the first 10 terms of the sequence

$$z_n = \frac{(-1)^n + 1}{n}.$$

Does the sequence converge or diverge? Why?

L'Hopital's Rule and Sequences

L'Hopital's rule can be helpful when computing the limit of sequences, but you must be careful. L'Hopital's rule applies to limits of differentiable functions on $\mathbb{R}$, but a sequence is a function on the natural numbers $\mathbb{N}$. The only time you can use L'Hopital's rule to determine the limit of a sequence $\{x_n\}$ is when

$$x_n = f(n)$$

for some differentiable function f. Thus, you can use it for the limit in application 8 but you cannot for the limit in application 11. If you apply L'Hopital's rule to a sequence, you should always justify its use.

2.3 The Squeeze Theorem

There are a variety of theorems that make the computing of limits (as well as the proofs) much easier. In what follows, we explore some of the basic theorems concerning the convergence of sequences. Again, many of these are almost identical to calculus theorems about limits.

Explore

Exploration 12. `IBLdynamics.com` Begin by using the application for the exercise to choose a random number. Next, create two different sequences that each converge to your random number. Call one of them $\{a_n\}$ and the other $\{b_n\}$ and enter the formulae for these sequences you've found into the boxes provided. Visually confirm that the sequences converge to the value you were given.

Exploration 13. Create a third sequence c_n as follows. Click the button on this website activity to generate a random weight value w. Then, let

$$c_n = wa_n + (1 - w)b_n.$$

What can you say about the relationship of the values c_n relative to the values of a_n and b_n?

Exploration 14. `IBLdynamics.com` Plot all three sequences on the same set of axes. Does this picture agree with your answers to exploration 13?

Conjecture

Conjecture 15. Based on your work above, complete the following statement of the Squeeze Theorem.

> If $a_n \leq c_n \leq b_n$ for all n and _____ then _____ .

Apply

Application 16. Use the Squeeze Theorem to compute the following limits.

1. $\displaystyle \lim_{n \to \infty} \frac{\cos n}{n}$

2. $\displaystyle \lim_{n \to \infty} \frac{n^2 + \sin n \cos n}{n^2}$

Application 17. Use the Squeeze Theorem to show that the following sequences converge.

1. $x_n = \dfrac{\cos(n^2)}{n^4}$

2. $x_n = \dfrac{e^{\sin(n)}}{n^2}$

3. $x_n = \dfrac{(-1)^n}{n}$

Follow the steps below to prove that

$$\lim_{n\to\infty} n^{1/n} = 1. \tag{2.2}$$

Application 18. Let $s_n = n^{1/n} - 1$. To what value should the sequence $\{s_n\}$ converge in order to prove equation 2.2?

Application 19. Manipulate the formula for s_n in application 18 to show that

$$n = (s_n + 1)^n.$$

Application 20. Look up the Binomial Theorem if you don't already know it. It tells you how to expand $(a + b)^n$. Use this to write out at least the first 3 terms of your work in application 19.

Application 21. Explain why the result of application 20 implies that

$$n > \frac{n(n-1)}{2} s_n^2.$$

Application 22. Solve the inequality in application 21 for s_n.

Application 23. Explain why $0 \le s_n$ for all n.

Application 24. Apply the Squeeze Theorem to prove the result in equation 2.2.

Prove

Proof 25. Use Definition 2.2 to prove the Squeeze Theorem as stated below.

Theorem 2.1 *If $a_n \le c_n \le b_n$ for all n and*

$$\lim_{n\to\infty} a_n = \lim_{n\to\infty} a_b = L$$

then

$$\lim_{n\to\infty} c_n = L.$$

2.4   Arithmetic Limit Theorems

All of the standard arithmetic operations can be done with sequences to create new sequences. In other words, sequences can be added, subtracted, multiplied, etc. There are theorems about the convergence of these new sequences in terms of the convergence of the original sequences. In what follows, suppose that both $\{a_n\}$ and $\{b_n\}$ are convergent sequences with

$$\lim_{n \to \infty} a_n = A$$

and

$$\lim_{n \to \infty} b_n = B.$$

Conjecture

Conjecture 26. Write a theorem about the convergence of the sequence $\{c_n\}$ if $c_n = a_n \pm b_n$.

Conjecture 27. Write a theorem about the convergence of the sequence $\{c_n\}$ if $c_n = K a_n$ where K is a constant.

Conjecture 28. Write a theorem about the convergence of the sequence $\{c_n\}$ if $c_n = a_n b_n$.

Conjecture 29. Write a theorem about the convergence of the sequence $\{c_n\}$ if $c_n = 1/b_n$. **Be careful with your hypothesis!**

Conjecture 30. Write a theorem about the convergence of the sequence $\{c_n\}$ if $c_n = a_n/b_n$. **Be careful with your hypothesis!**

Apply

Application 31. Find the values to which the following sequences converge. Do these results agree with your conjectures 26 through 30?

1. $a_n = \dfrac{3n^2 - 1}{10n + 5n^2}$

2. $b_n = \dfrac{\ln(n+1)}{\ln(1+4n)}$

3. $c_n = e^{-n}$

Application 32. Assume $\{a_n\}$ and $\{b_n\}$ are both real sequences. Provide counterexamples to each of the following statements.

1. If $\lim_{n\to\infty} a_n = 0$, then $\lim_{n\to\infty}(a_n b_n) = 0$.

2. If $\lim_{n\to\infty}(a_n b_n) = 0$, then either $\lim_{n\to\infty} a_n = 0$ or $\lim_{n\to\infty} b_n = 0$.

3. If $\lim_{n\to\infty}(a_n + b_n)$ does not exist, then either $\{a_n\}$ diverges or $\{b_n\}$ diverges.

4. If $\lim_{n\to\infty} \frac{a_{n+1}}{a_n} = 1$, then $\{a_n\}$ converges.

Do these problems suggest revising any of your conjectures 26 through 30?

Application 33. Recall that the Fibonacci sequence is given by $F_0 = 1$, $F_1 = 1$ and $F_n = F_{n-1} + F_{n-2}$. It can be shown that the sequence

$$\left\{ \frac{F_n}{F_{n-1}} \right\}$$

converges to some finite value ϕ. Assume that the sequence converges to prove that $\phi = \frac{1+\sqrt{5}}{2}$. This constant is known as the **golden ratio**.

Prove

Proof 34. Prove each of the statements in conjectures 26 through 30.

2.5 Bounded and Unbounded Sequences

Sequences that do not exceed some maximum value have some useful convergence properties that we will use in our study of dynamical systems. Thus, it is helpful to explore the convergence properties of such sequences. As usual, we begin with some definitions.

Definition 2.3 *A sequence $\{x_n\}$ is **bounded above** if there exists M such that $x_n \leq M$ for all n. A sequence $\{x_n\}$ is **bounded below** if there exists m such that $x_n \geq m$ for all n. A sequence that is both bounded above and bounded below is said to be **bounded**.*

Definition 2.4 *A sequence* $\{x_n\}$ *is said to have an* **infinite limit** *if for every* $M > 0$ *there exists* $N > 0$ *such that* $n > N$ *implies* $x_n > M$. *We denote this by*

$$\lim_{n \to \infty} x_n = \infty.$$

Definition 2.5 *A sequence* $\{x_n\}$ *is* **nondecreasing** *if* $x_{n+1} \geq x_n$ *for all* n. *A sequence* $\{x_n\}$ *is* **nonincreasing** *if* $x_{n+1} \leq x_n$ *for all* n. *We often refer to sequences having either of these properties as* **monotone sequences.**

A Note on Monotone Terminology

Mathematicians are sometimes a little careless with terminology as it relates to monotone sequences. We sometimes call a sequence monotone increasing even when it is technically monotone nondecreasing. That is, we might call a sequence where $x_{n+1} \geq x_n$ monotone increasing even though two consecutive terms of the sequence might be equal. I suppose the only reason for this is that it is a lot easier to say "increasing" than "non-decreasing."

Explore

Exploration 35. List some techniques that you might use to show that a sequence is monotone?

Exploration 36. Construct several different examples of monotone increasing sequences that are bounded above. Repeat with monotone decreasing sequences. What do you notice about each of these sequences?

Conjecture

Conjecture 37. Based on your work in exploration 36, conjecture a theorem about the limit of a monotone increasing sequence that is bounded above. State this conjecture in formal mathematical language. Repeat this for a monotone decreasing sequence bounded below.

Apply

Application 38. Use definition 2.4 to prove that

$$\lim_{n \to \infty} \frac{n^2 + 3}{n - 1} = \infty.$$

Application 39. Use definition 2.4 to prove that

$$\lim_{n \to \infty} \frac{n^2 + 3}{n + 1} = \infty.$$

Be careful! The proof is **not** identical to the proof of application 38.

Application 40. Let $x_1 = 1$ and define

$$x_{n+1} = \frac{n}{n+1} x_n^2.$$

1. Compute the terms x_1 through x_4.
2. Show that the sequence $\{x_n\}$ is monotone decreasing and bounded below. Conclude that the sequence converges.
3. **Challenge:** Prove that the sequence $\{x_n\}$ converges to 0.

Application 41. Let $x_1 = 1$ and define

$$x_{n+1} = \frac{1}{3}(x_n + 1).$$

Use the techniques of this section to prove that the sequence $\{x_n\}$ converges. **Challenge:** What does it converge to and why?

Application 42. Is the sequence $a_n = \dfrac{13^n}{n!}$ monotonic?

Application 43. If a sequence of positive real numbers is strictly decreasing, must it converge to zero?

Application 44. Can a monotone sequence which is unbounded converge?

Application 45. Can a bounded sequence which is not monotone converge?

Prove

At this point, you should be fairly comfortable with monotone sequences and hopefully you conjectured that a monotone increasing sequence that is bounded above converges. The formal statement of that theorem is as follows.

Theorem 2.2 *If $\{x_n\}$ is a monotone increasing sequence bounded above then $\{x_n\}$ converges. Similarly, if $\{x_n\}$ is a monotone decreasing sequence bounded below then $\{x_n\}$ converges.*

Before turning to the proof of this theorem, we need to discuss an important property of the real numbers called the **Completeness Axiom.**

Axiom 2.1 (Completeness Axiom) *If $S \subset \mathbb{R}$ is bounded above, then S has a least upper bound called a* **supremum** *and denoted* $\sup S$. *Similarly, if S is bounded below, it must have a greatest lower bound called an* **infimum** *and denoted* $\inf S$.

Infimums and supremums are like minimums and maximums except they do not necessarily need to be contained in the set S.

Exploration 46. Find the infimums and supremums of the following sets if they exist. Determine, if possible, whether each is included in the given set or not.

1. $S = (0, 1)$
2. $S = [0, 1]$
3. $S = \{x \in (0, 1) \mid x \in \mathbb{Q}\}$
4. $S = \{x \in \mathbb{R} \mid x = 1/n,\ n \in \mathbb{Z}\}$

Notes on the Completeness Axiom

It is important to understand that the Completeness Axiom is an axiom and not a theorem. That means that this property is one of the fundamental (and hence non-provable) assumptions used to define the real numbers. It is akin to one of the axioms of Euclidean geometry such as "through any 2 distinct points there is one and only one line." All of geometry is built up from these axioms. Similarly, all of the properties of the real numbers are built from a set of axioms one of which is Completeness.

Also note that the Completeness Axiom applies to *sets* of real numbers, not just sequences. However, a sequence is a set if we ignore the ordering.

With this background in mind, we can proceed with the proof of this theorem. You will prove it for monotone increasing sequences with the proof for monotone decreasing sequences being similar.

Proof 47. Let $\{x_n\}_{n=1}^{\infty}$ be a monotone increasing sequence bounded above. Explain why the set $\{x_n\}$ has a supremum u. Let $\sup\{x_n\} = u$.

Our goal is to show that in fact the sequence $\{x_n\}$ converges to u.

Proof 48. Fix $\varepsilon > 0$ and explain why $u - \varepsilon$ is not a supremum of $\{x_n\}$. Use this to conclude that there exists $N > 0$ such that if $n > N$ then $x_n > u - \varepsilon$.

Proof 49. Conclude the proof by using the result of proof 48 to show that $|x_n - u| < \varepsilon$ for all $n > N$ and thus

$$\lim_{n \to \infty} x_n = u.$$

2.6 Subsequences

It is sometimes important to consider only a selection of terms in a sequence, rather than all of the terms. For example, you might want to look at the sequence of even indexed terms or the sequence of odd indexed terms because doing so makes it easy to determine whether the sequence converges or diverges. A new sequence taken from a given sequence is called a subsequence and is formally defined below.

Definition 2.6 *Given a sequence* $\{x_n\}_{n=1}^{\infty}$ *and an infinite set of natural numbers* $v_1 < v_2 < v_3 < \cdots$, *a* **subsequence** $\{\hat{x}_k\}_{k=1}^{\infty}$ *is the sequence where* $\hat{x}_k = x_{v_k}$.

Explore

Exploration 50. Let's start with a basic example to illustrate this concept. Consider the sequence

$$\{x_n\}_{n=1}^{\infty} = \left\{\frac{(-1)^n}{n}\right\}_{n=1}^{\infty}.$$

1. What is the subsequence $\{a_k\} = \{x_{2k}\}$?
2. List properties of this subsequence.
3. What does the subsequence $\{a_k\}$ converge to?

4. What is the subsequence $\{b_k\} = \{x_{2k-1}\}$?

5. List properties of this subsequence.

6. What does the subsequence $\{b_k\}$ converge to?

7. What does this tell you about the original sequence? What sequence theorem are you using here?

Exploration 51. In applications 6–9, you computed the limits of various sequences. For each of these sequences, perform the following:

1. Construct a subsequence that converges to the same value as the original sequence.

2. Construct a subsequence that diverges, or show one does not exist.

Exploration 52. Consider the sequence $x_n = \cos(\frac{n\pi}{2})$

1. Does this sequence converge or diverge?

2. Does the subsequence $\{x_{2n+1}\}$ converge or diverge?

3. Does the subsequence $\{x_{4n}\}$ converge or diverge?

Exploration 53. Construct an unbounded sequence with a bounded subsequence.

Exploration 54. Consider the sequence $x_n = (-1)^n$. Find two disjoint subsequences that converge to -1 and 1, respectively.

Prove

Proof 55. Prove that, if a sequence is convergent, its subsequences are also convergent, and moreover, they converge to the same value.

2.7 Liminfs and Limsups

In proving theorem 2.2, you used either the supremum or the infimum concept to show that a bounded monotone sequence converges. In this section, we will take this type of analysis one step further by creating sequences of these quantities and then computing their corresponding limits.

Let $\{x_n\}$ be a sequence of real numbers. Define

$$u_N = \inf\{x_n : n > N\} \tag{2.3}$$

and

$$v_N = \sup\{x_n : n > N\}. \tag{2.4}$$

Carefully note how both n and N are used in these definitions. For a given u_N or v_N, the number N is fixed while n indicates all of the subscripts greater than this fixed value. Thus, each u_N is an infimum of the set that is composed of all elements of the sequence $\{x_n\}$ that have an index greater than N. You might say that each u_N is an infimum of the "tail" of the original sequence. Similarly, with v_N except that these are supremums.

These u_N and v_N values form sequences in their own right. We can ask whether these new sequences converge and, if so, what are the limits. Additionally, it is natural to ask how these limits do, or do not, relate to the limit of the original sequence when one exists.

Definition 2.7 *The limit infimum of a sequence $\{x_n\}$ (denoted $\liminf$) is*

$$\liminf x_n = \lim_{N \to \infty} u_N = \lim_{N \to \infty} \inf\{x_n : n > N\}. \tag{2.5}$$

Similarly,

$$\limsup x_n = \lim_{N \to \infty} v_N = \lim_{N \to \infty} \sup\{x_n : n > N\}. \tag{2.6}$$

Explore

Exploration 56. IBLdynamics.com Use the sequence $\{x_n\}$ with

$$x_n = 1 + \frac{(-1)^n}{n}$$

to answer the following questions.

1. Make a list of the sequence elements x_1 through x_6. Plot these points on a graph with n on the horizontal axis and x_n on the vertical axis.

2. Let's start with the infimums. Compute u_1 through u_6 where

$$
\begin{aligned}
u_1 &= \inf\{x_n : n > 1\} \\
u_2 &= \inf\{x_n : n > 2\} \\
u_3 &= \inf\{x_n : n > 3\}
\end{aligned}
$$

etc. Remember, an infimum can be a minimum!

3. Now, let's turn to the supremums. Compute v_1 through v_6 where

$$\begin{aligned} v_1 &= \sup\{x_n : n > 1\} \\ v_2 &= \sup\{x_n : n > 2\} \\ v_3 &= \sup\{x_n : n > 3\} \end{aligned}$$

etc. Remember, a supremum can be a maximum!

4. What are some of the things you notice about these two sequences?

   (a) Are they monotone or not?

   (b) Are they subsequences of the original sequence?

   (c) Do either appear to converge?

   (d) If so, to what?

5. The tool on the website for exploration 56 shows the points of the sequence x_n, along with u_N and v_N. Does this agree with what you observed in answering the preceding questions? Explain.

Exploration 57. ⬚ IBLdynamics.com ⬚ Repeat the activities you did in exploration 56 but with the sequence

$$x_n = (-1)^n \left(1 + \frac{1}{n} \right).$$

Again, there is a tool on the website that you should use to verify your thinking.

Exploration 58. ⬚ IBLdynamics.com ⬚ There is one final tool on the website that generates a random sequence $\{x_n\}$ and again shows the sequences $\{u_N\}$ (green dots) and v_N (red dots). What do these dots (i.e., sequences) do as you scroll through?

Exploration 59. Give an example of a sequence $\{x_n\}$ where either the sequence $\{u_N\}$ or the sequence $\{v_N\}$ is *not* a subsequence of $\{x_n\}$.

Conjecture

Let's see if we can turn these observations about specific examples into formal mathematical statements.

Conjecture 60.

> Given any sequence $\{x_n\}$ (convergent or not), the sequence of infimums $\{u_N\}$ is ——————— . Similarly, the sequence of infimums $\{v_N\}$ is ——————— .

Conjecture 61.

> If $\{x_n\}$ is a bounded sequence, then the sequence $\{u_N\}$ is ——————— and the sequence $\{v_N\}$ is ——————— .

Conjecture 62.

> Suppose that the sequence $\{x_n\}$ converges to L. Then the sequence $\{u_N\}$ ——————— and the sequence $\{v_N\}$ ——————— .

Conjecture 63.

> Suppose that for a sequence $\{x_n\}$ the sequences $\{u_N\}$ and $\{v_N\}$ converge to the same value L. Then the sequence $\{x_n\}$ ——————— .

Conjecture 64. Combine conjectures 62 and 63 into a single theorem.

Apply

Application 65. Find the lim sup and lim inf of the sequence

$$a_n = n^{\sin\left(\frac{n\pi}{2}\right)}.$$

Application 66. Find the lim sup and lim inf of the sequence

$$a_n = \sin\left(\frac{n\pi}{3}\right).$$

Application 67. Find the lim sup and lim inf of the sequence

$$a_n = \begin{cases} 2^{\frac{1}{n+1}} & \text{if } n \text{ is even} \\ 1 & \text{if } n \text{ is odd} \end{cases}.$$

Prove

The examples that you have worked through in this section have led you to discover the important relationships between the lim sup and lim inf of a sequence and how these quantities relate to the limit of the given sequence. In particular, you conjectured the following theorem.

Theorem 2.3 *A sequence $\{x_n\}$ on $\mathbb{R}$ converges to a finite value L if and only if*

$$\liminf x_n = \limsup x_n = L.$$

We will approach the proof of this in several steps.

Lemma 2.1 *Given any sequence $\{x_n\}$, the sequence of infimums $\{u_n\}$ is monotone increasing. Similarly, the sequence of supremums $\{v_n\}$ is monotone decreasing.*

Proof 68. Prove Lemma 2.1.

Lemma 2.2 *If the sequence $\{x_n\}$ is bounded, then the sequence of infimums $\{u_n\}$ converges. Similarly, the sequence of supremums $\{v_n\}$ converges.*

Proof 69. Prove Lemma 2.2.

The following sequence of proofs will lead you through a proof of Theorem 2.3.

Proof 70. Let's begin by assuming that

$$\lim_{n \to \infty} x_n = L \qquad (2.7)$$

and prove that $\liminf x_n = \limsup x_n = L$.

1. Explain why we can apply Lemma 2.2 to conclude that both $\liminf x_n$ and $\limsup x_n$ exist.

2. Let $U = \liminf x_n$ and $V = \limsup x_n$. Explain why $U \leq V$.

3. Assume that $U \neq V$ and use equation 2.7 to reach a contradiction and thus prove this part of the theorem.

Now let's prove the other direction of Theorem 2.3. For the next two proofs, assume that

$$\liminf x_n = \limsup x_n = L.$$

Proof 71.

1. Prove that for every $\varepsilon > 0$ there exists N_0 such that

$$\sup\{x_n : n > N_0\} < L + \varepsilon.$$

2. Explain why this implies that $x_n < L + \varepsilon$ for all $n > N_0$.

Proof 72.

1. Prove that for every $\varepsilon > 0$ there exists N_1 such that

$$\inf\{x_n : n > N_1\} > L - \varepsilon$$

and explain why this implies that $x_n > L - \varepsilon$ for all $n > N_1$.

Proof 73. How should you choose N so that **both** proof 71 and 72 are true? Use this to explain why $|L - x_n| < \varepsilon$ for all $n > N$. Finish the proof of Theorem 2.3.

2.8 Cauchy Sequences

Up to this point, the only tools we have to determine whether a sequence converges or not is to either compute its limit or show it is a bounded monotone sequence. However, for many sequences it is difficult to compute the actual limit of the sequence directly, even though we may suspect that it converges to something. Additionally, there are convergent sequences that are not monotone such as the sequence

$$x_n = \frac{(-1)^n}{n}.$$

It would be helpful to have another tool that allows us to determine whether a sequence converges or not, but does not actually force us to compute the limit of the given sequence. This idea is not new to you. In calculus, you used a variety of tools to determine whether an infinite series converges or not (remember the ratio test). None of these tools actually computed the value of the infinite series. Additional work was needed to complete this task when that was even possible.

You have probably already noticed that if $\{x_n\}$ is a convergent sequence then as you compute terms with large n values, the distance between terms gets really small. This idea is summarized in the following definition.

Definition 2.8 *A sequence $\{x_n\}$ is a* **Cauchy sequence** *if for every $\varepsilon > 0$ there exists $N > 0$ such that if $m, n > N$, then $|x_m - x_n| < \varepsilon$.*

Explore

Exploration 74. Discuss the similarities and differences of definition 2.8 to that of convergence given in definition 2.2.

Explorations 75 through 78 illustrate a common mistake when working with Cauchy sequences. It seems reasonable to assume that if the distance between consecutive terms in the sequence goes to 0, then the sequence is a Cauchy sequence. This is not the case as the following set of exercises demonstrates.

Exploration 75. Show that the sequence $\left\{ \sqrt{n+1} - \sqrt{n} \right\}$ converges to 0.

Exploration 76. Show that sequence where $x_n = \sqrt{n}$ diverges.

Exploration 77. Show that for the sequence in exploration 76,

$$\lim_{n \to \infty} |x_{n+1} - x_n| = 0.$$

Exploration 78. Show that sequence in exploration 76 is *not* a Cauchy sequence by choosing $\varepsilon = 1$ and showing that for all $N > 0$, if $N < m = k^2$ and $N < n = (k+1)^2$ then

$$|x_m - x_n| = \varepsilon.$$

Exploration 79. What phrase in definition 2.8 allows us to choose a specific value of ε to show that the sequence $\{x_n\}$ is not a Cauchy sequence?

Conjecture

Conjecture 80. Suppose that $\{x_n\}$ is a convergent sequence. Do you think that it must also be a Cauchy sequence? Why or why not?

Conjecture 81. Suppose that $\{x_n\}$ is a Cauchy sequence on $\mathbb{R}$. Do you think that it must also converge to a number in $\mathbb{R}$? Why or why not?

Conjecture 82. Suppose that $\{x_n\}$ is a Cauchy sequence on the open interval $(0, 1)$. Do you think that it must also converge to a number in $(0, 1)$? Why or why not?

Conjecture 83. Suppose that $\{x_n\}$ is a Cauchy sequence on the closed interval $[0, 1]$. Do you think that it must also converge to a number in $[0, 1]$? Why or why not?

Conjecture 84. Suppose that $\{x_n\}$ is a Cauchy sequence where x_n a rational number for all n. Do you think that it must also converge to a rational number? Why or why not?

Apply

Application 85. Prove that the series $\sum_{n=1}^{\infty} \frac{1}{n^2}$ converges by showing that the sequence of partial sums is Cauchy.

Application 86. Prove that, if $\{a_n\}$ is Cauchy, then the sequence $\{a_n^2\}$ is also Cauchy. Show the converse is not true.

Application 87. Suppose $\{a_n\}$ is a Cauchy sequence and that there exists a subsequence of $\{a_n\}$ that converges to L. Prove that $\{a_n\}$ converges to L as well.

Prove

Before proceeding with a proof, a discussion of the *Conjecture* questions above is in order. They point towards an important analysis concept called completeness (that is different from the Completeness Axiom). Ideally, one wants every Cauchy sequence on some metric space X to converge to an element of X (a metric space is basically just something where there is a well-defined notion of distance). Conjectures 82 and 84 demonstrate that this is not necessarily true. It is true for the real numbers and the closed interval $[0, 1]$, but not for the open interval $(0, 1)$ or the rational numbers. We say that a metric space X is **complete** if every Cauchy sequence in X converges to a point in X. We won't pursue the concept of completeness any further except to note that the real line $\mathbb{R}$ is complete and any closed interval $[a, b]$ is complete. This gives us the following theorem.

Theorem 2.4 *A sequence $\{x_n\}$ converges on $\mathbb{R}$ if and only if $\{x_n\}$ is a Cauchy sequence.*

As we have done in the past, we will guide you through the proof of this theorem.

Proof 88. One direction of this proof is fairly straightforward. Prove that if $\{x_n\}$ is a convergent sequence on $\mathbb{R}$ then $\{x_n\}$ is a Cauchy sequence.

We can now work on showing that if our sequence $\{x_n\}$ is Cauchy then it converges.

Proof 89. Prove that if $\{x_n\}$ is a Cauchy sequence on $\mathbb{R}$, then $\{x_n\}$ is bounded. **Hint:** What is the contrapositive statement?

Proof 90. Fix $\varepsilon > 0$ and use the fact that $\{x_n\}$ is a Cauchy sequence to show that $x_m + \varepsilon$ is an upper bound for x_n for all $n > N$.

Proof 91. Now show that

$$v_N = \sup\{x_n : n > N\} \leq x_m + \varepsilon$$

and conclude that $v_N - \varepsilon \leq x_m$ for all $m > N$.

Proof 92. Explain why $v_N - \varepsilon \leq x_m$ for all $m > N$ implies that $v_N - \varepsilon$ is a *lower bound* for $\{x_m : m > N\}$. Conclude that

$$v_N - \varepsilon < \inf\{x_m : m > N\} = u_N.$$

Proof 93. Explain each of the following inequalities.

$$\limsup x_n \leq v_N \leq u_N + \varepsilon \leq \liminf x_n + \varepsilon$$

Proof 94. Explain why

$$\limsup x_n \leq \liminf x_n$$

and

$$\liminf x_n \leq \limsup x_n.$$

Be careful with the subscripts!

Proof 95. Conclude that $\liminf x_n = \limsup x_n$ and explain why this implies that the original sequence converges.

3

Fixed Points and Periodic Points

3.1 Fixed Points

The study of almost every dynamical systems begins with determining the values of the fixed points. Fixed points were introduced in section 1.2 where you learned that a fixed point $\tilde{x}$ satisfies

$$f(\tilde{x}) = \tilde{x}. \tag{3.1}$$

This equation tells us that fixed points can be found algebraically by solving the equation $f(x) = x$ for x. We call this special value $\tilde{x}$. This is often a fairly straightforward process. For example, it is easy to see that the fixed points of $f(x) = x^2$ are $x = 0$ and $x = 1$ since these are the only solutions to $x^2 = x$. Sometimes, we are unable to compute the exact values of fixed points. Can you algebraically find the fixed points of $\cos x$? This definition also tells us something important about the graphical nature of fixed points. It says that fixed points are the points of intersection of the graphs of $y = f(x)$ and the diagonal line $y = x$. You saw this when you were doing the graphical analysis problems in chapter 1. Figure 3.1 illustrates this with the two functions discussed above.

Once we know the values of a fixed point, the next thing we need to determine is its relationship to other orbits of the dynamical system. In particular, do the orbits of nearby starting values converge to the fixed point, go away from the fixed point, or exhibit some other behavior less easily categorized?

Definition 3.1 *A fixed point $\tilde{x}$ is* **attracting** *(or* **stable***) if there exists an open interval U containing $\tilde{x}$ such that if $x_0 \in U$, then*

$$\lim_{n \to \infty} f^n(x_0) = \tilde{x}.$$

In short, a fixed point is attracting if nearby initial conditions converge to it under iteration. There are also **repelling** (or **unstable**) fixed points and while the formal definition is a bit more complicated, the basic premise is that

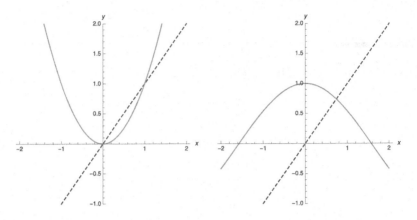

FIGURE 3.1
Fixed points are the intersection points of the graphs of $y = f(x)$ and $y = x$.
(Left) The function $f(x) = x^2$ has two fixed points. (Right) The function
$f(x) = \cos x$ has a single fixed point.

nearby initial conditions move away from the fixed point under iteration. We
will formally define this concept later. Finally, fixed points that are neither
attracting nor repelling are called **neutral.**

3.1.1   Fixed Points of Linear Systems

We begin the exploration into fixed points by investigating the dynamics of
linear dynamical systems of the form

$$x_{n+1} = mx_n + b \tag{3.2}$$

where m and b are fixed parameters.

Explore

Exploration 1. Find the values of the fixed point(s) in terms of the parameters m and b. Are there any values that either of these parameters cannot
take?

Exploration 2. ⎸`IBLdynamics.com`⎸ Use the tool on the website to explore
the dynamics of linear systems when $b = 0$. You should pay particular attention to how the value of m affects the fixed point and/or its stability. Describe
what you see.

Exploration 3. $\boxed{\text{IBLdynamics.com}}$ Use the tool on the website to explore the dynamics of linear systems for other values of b.

1. Describe how the value of m affects the fixed point and/or its stability. Is this different from what you observed in the previous exercise?

2. How does b change the fixed point and/or its stability?

3. Does the stability of the fixed point depend in any way on the initial condition?

Conjecture

Conjecture 4. The work in the previous two problems suggests a theorem about linear dynamical systems. Fill in the blanks below to complete it.

> Consider the linear dynamical system
>
> $$x_{n+1} = mx_n + b.$$
>
> If _____, then there is a unique fixed point at _____. If _____, then the fixed point is attracting. If _____, then the fixed point is repelling.

Apply

In this section, you will model and analyze the population of walleye in Teelo Lake where there is a strict, fixed limit on the total number of fish caught per year. Let n measure the number of years since 2020 (i.e., when $n = 10$ it is 2030) and let w_n denote the walleye population at year n.

Application 5. Model the walleye population assuming that there is no fishing whatsoever. Assume that the rate of change of the walleye population from year n to year $n + 1$ is proportional to the walleye population at year n with proportionality constant $r > 0$. Show that

$$w_{n+1} = (1 + r)w_n.$$

Application 6. What are the conditions on r that guarantee that the walleye population will not go extinct in Teelo Lake? Describe the dynamics of the walleye population in this scenario.

Application 7. Suppose that exactly h fish are caught per year. Modify equation 5 to model this situation. In this harvesting model, what is the equilibrium walleye population and for what values of r and h (if any) is it attracting?

Application 8. Are there situations where the walleye population will go extinct? Describe them in terms of the model parameters and the initial conditions.

Application 9. Consider a different harvesting model. Instead of assuming a fixed number of fish caught per year, assume that the number of fish caught in year $n+1$ is proportional to the walleye population of year n. Let h denote the constant of proportionality. What is the model equation in this case?

Application 10. For the model in application 9, what is the equilibrium walleye population and for what values of r and h (if any) is it attracting? Are there situations where the walleye population will go extinct? Describe them in terms of the model parameters and the initial conditions.

Application 11. Discuss the pros and cons of these two models.

Prove

In this section, you will prove the theorem about fixed points of linear dynamical systems from conjecture 4.

Theorem 3.1 *Consider the linear dynamical system*

$$x_{n+1} = mx_n + b.$$

If $m \neq 1$ then there is a unique fixed point at $x = b/(1-m)$. If $|m| < 1$ then the fixed point is attracting. If $|m| > 1$ then the fixed point is repelling.

Proof 12. Use mathematical induction to prove that the solution to equation 3.2 with $b = 0$ and initial condition x_0 is $x_n = m^n x_0$.

Proof 13. Prove that the fixed point of equation 3.2 when $b = 0$ is attracting if $|m| < 1$ and repelling if $|m| > 1$.

Proof 14. Assume that $m \neq 1$ and let

$$y_n = x_n - \frac{b}{1-m}.$$

Rewrite equation 3.2 in terms of y_n and y_{n+1}.

Proof 15. Use the results of proof 13 to prove that the fixed point of equation 3.2 is attracting if $|m| < 1$ and repelling if $|m| > 1$.

3.1.2 Attracting Fixed Points of Nonlinear Systems

We've now completely described the stability of fixed points in linear dynamical systems on $\mathbb{R}$. Although the dynamics of nonlinear systems can be much more complicated, the insights gained from studying fixed points of linear systems is valuable. Keep these insights in mind as we turn our attention to analyzing fixed points of non-linear systems.

Consider the dynamical system

$$x_{n+1} = f(x_n) \tag{3.3}$$

where we assume that $f : \mathbb{R} \to \mathbb{R}$ is a differentiable function with a continuous derivative.

Your goal in this section is almost identical to what you did in subsection 3.1.1; find conditions on the function f that determine the stability of a fixed point.

Almost every exploration in this book from this point on will focus on the dynamics of the *logistic family* of functions

$$f_a(x) = ax(1 - x) \tag{3.4}$$

and the associated dynamical system

$$x_{n+1} = f_a(x_n) = ax_n(1 - x_n). \tag{3.5}$$

This is another population dynamics model. In the **Apply** exercises below, you will be guided through the development of this model. We choose to use this dynamical system to demonstrate almost every concept that follows because

- the algebra and calculus required for most calculations is straightforward,

- the dynamics can be readily interpreted in a modeling context, and most importantly,

- the lessons learned by studying it are transferable to almost every other nonlinear dynamical system on $\mathbb{R}$.

Explore

We begin with a few exercises that highlight the graphical properties of the logistic family of functions of equation 3.4. Knowing these properties will be essential when you begin exploring the dynamics.

Exploration 16. Plot the graph of $y = f_a(x)$ for $a > 0$ and answer the following questions.

1. What are the x-intercepts of $y = f_a(x)$? Do they depend on the value of a?

2. What is the critical point of $y = f_a(x)$? Does it depend on the value of a? Is it a maximum or a minimum?

3. What is the y-value of the critical point (this is sometimes called a *critical value*)? Does it depend on the value of a?

With those fundamentals in hand, we can begin exploring the dynamics of equation 3.5.

Exploration 17. Show that if $a \neq 0$ then the fixed points of equation 3.5 are $x = 0$ and $x = (a - 1)/a$.

Exploration 18. $\boxed{\texttt{IBLdynamics.com}}$ Use the tool on the website to determine the values of a for which the fixed point at $x = 0$ is attracting, repelling, or neither. Pay attention to the relationship between the graphs of $y = f_a(x)$ and $y = x$ at the fixed point $x = 0$. What is this relationship when the fixed point is attracting versus repelling?

Exploration 19. $\boxed{\texttt{IBLdynamics.com}}$ Use the tool on the website to determine the values of a for which the fixed point at $x = (a - 1)/a$ is attracting, repelling, or neither. Pay attention to the relationship between the graphs of $y = f_a(x)$ and $y = x$ at the fixed point $x = (a - 1)/a$. What is this relationship when the fixed point is attracting versus repelling? This is harder to see than the last problem! For some parameter values where the fixed point is attracting, you will see almost the exact same relationship as you did in exploration 18. But for others, the relationship will be harder to see. Here is a clue if you struggle to see this: What is the "slope" of $f_a(x)$ at the fixed point when the fixed point is just to the left of the critical point? What is the "slope" of $f_a(x)$ at the fixed point when the fixed point is just to the right of the critical point?

Just to convince you that what you just observed is not totally dependent on the choice of function, let's repeat these exercises with a different family of functions.

Exploration 20. $\boxed{\texttt{IBLdynamics.com}}$ For the family of functions $s_a(x) = a\sin(x)$, use the sine iteration tool on the website to determine the values of a where the fixed point at $x = 0$ is attracting, repelling, or neither. What is this relationship between the two graphs shown on the website near $x = 0$ when the fixed point is attracting versus when it is repelling?

Exploration 21. Think about the explorations above and what you observed about the stability of the fixed points. Reflect on the relationship between the graph of the dynamical function and the diagonal line. Is there a computable quantity that determines whether a fixed point is attracting or repelling?

Conjecture

Conjecture 22. Now let's put this together as a formal mathematical statement.

Assume that $f(\tilde{x}) = \tilde{x}$ and that f is differentiable with a continuous derivative in a neighborhood of $\tilde{x}$. If _____ then $\tilde{x}$ is an attracting fixed point.

Apply

Now seems a good point to discuss why the logistic model in equation 3.5 is a reasonable population dynamics model. Recall from section 1.4 that a general form of many population models is

$$p_{n+1} = \lambda(p_n)V(p_n)p_n, \qquad (3.6)$$

where n denotes the generation, p_n is the population at generation n, λ is the reproductive rate function, and V is the viability function that measures the survival percentage of newborns.

Application 23. Let's begin by assuming that $\lambda(p) = a$. What is the population dynamics interpretation of this assumption?

Application 24. Let's turn to the viability function and suppose that

$$V(p) = 1 - \frac{p}{K}.$$

1. What is $V(0)$ and $V(K)$? What is the interpretation of these properties?

2. More generally, what can you say about the value of $V(p)$ for $0 \le p \le K$? Why is this important?

3. Is $V(p)$ increasing or decreasing? What is the interpretation of this property?

4. The parameter K is a population size. What does it signify in this model?

5. If we assume that $K = 1$, then how can we interpret the variable p?

Application 25. Show that the assumptions of the previous two exercises give the logistic model of equation 3.5.

Now let's put together the things we've learned about fixed point stability to analyze the fixed points of this model.

Application 26. For equation 3.5, use conjecture 22 to determine the values of a where the fixed point at $x = 0$ is attracting. Your answer should be an interval. Interpret this in terms of population dynamics.

Application 27. For equation 3.5, use conjecture 22 to determine the values of a where the fixed point at $x = (a - 1)/a$ is attracting. Your answer should be an interval. Interpret this in terms of population dynamics.

Application 28. Let's tie things together by considering the fixed point at $x = 0$ of $f_a(x) = a \sin x$ from exploration 20. Determine the values of a where this fixed point is attracting.

Prove

Theorem 3.2 (Attracting Fixed Point Theorem) *Assume that $f(\tilde{x}) = \tilde{x}$ and that f is differentiable with a continuous derivative in a neighborhood of $\tilde{x}$. If $|f'(\tilde{x})| < 1$ then $\tilde{x}$ is an attracting fixed point.*

The following exercises lead you to one proof of this theorem. There are other approaches to proving this fundamental result and we encourage you to think about other methods of proof.

In this proof we assume, without loss of generality, that the fixed point is $x = 0$. If it is not, we would simply change variables in a manner similar to proof 14 in section 3.1.1.

Let's begin by looking at the case where $0 \leq f'(0) < 1$.

Proof 29. Show that there exists $\varepsilon > 0$ such that if $0 < x < \varepsilon$ then $0 \leq f(x) < x$.

Proof 30. Show that if $0 < x_0 < \varepsilon$ and $x_{n+1} = f(x_n)$, then the sequence $\{x_n\}$ is decreasing and bounded below. Cite the reason why this implies that the sequence converges.

Proof 31. Show that sequence $\{x_n\}$ converges to 0 by

1. proving that it cannot converge to a value of x strictly less than 0, and then

2. assuming that it converges to a value of x strictly greater than 0 and then reaching a contradiction.

Proof 32. Repeat exercises 29 through 31 modified appropriately for $-\varepsilon < x < 0$ to complete the proof.

With this done, the case where $-1 < f'(x) < 0$ is straightforward.

Proof 33. Let $g(x) = -f(x)$. What can you say about $g'(0)$? Why does the work above imply that $g(x)$ has an attracting fixed point at $x = 0$? Why does this imply that $f(x)$ has an attracting fixed point at $x = 0$?

3.1.3 Repelling Fixed Points of Nonlinear Systems

In definition 3.1, we defined an attracting fixed point but delayed the definition of repelling fixed point. So let us define that now.

Definition 3.2 *A fixed point $\tilde{x}$ of $f(x)$ is* **repelling** *(or* **unstable***) if there exists an open interval U containing $\tilde{x}$ such that for all $x_0 \in U$, other than $\tilde{x}$, there exists $N > 0$ such that $f^N(x_0) \notin U$.*

This definition, while a bit complicated sounding, simply says that a fixed point is repelling if nearby points iterate away from the fixed point. More precisely, it says that if we start off at any point x_0 in some small neighborhood U of the fixed point, then at some time (the Nth iterate) the orbit of x_0 leaves the interval U. Note that it does not say that the orbit leaves and never comes back. This can happen. We will see instances of this type of behavior later.

Explore

Your goal in this section is almost identical to what you did in section 3.1.2; to determine conditions that insure that a fixed point is either attracting or repelling.

Exploration 34. $\boxed{\text{IBLdynamics.com}}$ Review the explorations from section 3.1.2 and note when each of the fixed points is repelling. On the website this is broken up into two tools, one for the logistic function, and one for the sine function.

Conjecture

Conjecture 35. Reflecting on both the explorations and Theorem 3.2 complete the following to conjecture a theorem on repelling fixed points.

Assume that $f(\tilde{x}) = \tilde{x}$ and that f is differentiable with a continuous derivative in a neighborhood of $\tilde{x}$. If _____, then $\tilde{x}$ is a repelling fixed point.

Apply

Application 36. Return to the logistic equation (3.5) and note the interval of a-values for which the fixed point at $x = 0$ is attracting. Now determine the interval of a-values where the fixed point at $x = 0$ is repelling.

Application 37. Again for the logistic equation, review the interval of values of a where the fixed point at $x = (a - 1)/a$ is attracting. Determine the interval(s) of a-values where it is repelling. Restrict your work to the case where $a > 0$.

Application 38. | IBLdynamics.com | Construct a fixed point graph for the logistic function. This is a graph with the parameter a on the horizontal axis and the location of the fixed points on the vertical axis. There will be two curves on this graph. Indicate the stability of each fixed point by coloring or dashing each curve (for example, you might color attracting fixed points aquamarine and repelling fixed points ruby).

1. Remember that the logistic equation models population dynamics. Let's interpret the fixed point graph that you just created in this context.

   (a) For what values of a will the population go extinct?
   (b) For what values of a will the population approach a non-zero equilibrium state?
   (c) Do all initial conditions $x_0 \in [0,1]$ approach this equilibrium population? Why or why not?
   (d) How do the equilibrium values change as a function of the parameter a?

2. Let's look at some other mathematical features of this fixed point graph. What is the relationship between the two curves on the fixed point graph at the value(s) of a where the stability changes from attracting to repelling or vice versa?

3. Go to the tool on the website for this exercise. Describe what happens to the graph of $f(x) = ax(1 - x)$ as you increase a? Is this consistent with your observations in exploration 16?

4. Now, look at the fixed point at $x = 0$ and manipulate the graph through the value of a where this point changes from being attracting to being repelling. Describe what happens both in terms of the graph and in terms of both fixed points. Write a few sentences to relate this behavior to the fixed point graph that you created above.

Application 39. If you need any additional practice on finding fixed points and determining their stability, here are few more problems. Find all of the fixed points of the following functions and determine whether each is attracting or repelling.

1. $f(x) = x^2 - 1/8$
2. $f(x) = x^2 - 2$
3. $f(x) = e^x - 1$
4. $f(x) = x^3$
5. $f(x) = \sqrt{x}$

Prove

Theorem 3.3 (Repelling Fixed Point Theorem) *Assume that $f(\tilde{x}) = \tilde{x}$ and that f is differentiable with a continuous derivative in a neighborhood of $\tilde{x}$. If $|f'(\tilde{x})| > 1$ then $\tilde{x}$ is a repelling fixed point.*

The following exercises lead you to one proof of this theorem. As in the attracting fixed point proof, we assume without loss of generality that the fixed point is $x = 0$ and begin by considering the case where $f'(0) > 1$.

Proof 40. Show that there exists $\varepsilon > 0$ such that if $0 < x < \varepsilon$, then $x < f(x)$.

Proof 41. Show that if $0 < x_n < \varepsilon$ and $x_{n+1} = f(x_n)$, then $x_{n+1} > x_n$.

Proof 42. Show that if $0 < x_n < \varepsilon$, then there exists $N > 0$ such that $x_N > \varepsilon$.

Proof 43. Repeat proofs 40 through 42, modified appropriately, for $-\varepsilon < x < 0$ to complete the proof.

Proof 44. Now prove the theorem assuming that $f'(0) < -1$ by modifying the argument of proof 33.

3.1.4   Neutral Fixed Points of Nonlinear Systems

We now turn our attention to analyzing so-called neutral fixed points of non-linear systems.

Definition 3.3 *A fixed point $\tilde{x}$ of $f(x)$ is* **neutral** *if $|f'(\tilde{x})| = 1$.*

As you will begin to see in the explorations below, neutral fixed points are important when thinking about families of dynamical systems. This is because when a fixed point is neutral, a small change to the function being iterated can lead to qualitatively different types of dynamics. This change in behavior is called a bifurcation and will be the subject of chapter 5.

Our immediate goal is not to explore bifurcations, but simply to categorize some of the dynamics that occur in the neighborhood of a neutral fixed point. The dynamical ideas presented here are not a complete list of the properties of neutral fixed points. Instead, we focus on the most common dynamical behaviors that can occur. These are easily understood through a combination of calculus ideas and graphical analysis.

Explore

Exploration 45. In each of the graphs of figure 3.2, $x = 0$ is the only fixed point and is a neutral fixed point. What is the value of $f'(0)$ in each? Use graphical analysis to iterate the function and describe orbits with initial conditions on either side of the fixed point. A version of this figure is available on the website if you wish to print out a copy for doing graphical analysis.

Exploration 46. Describe the concavity of each of the functions shown above. What quantity determines concavity?

Exploration 47. Pretend that you could reach into the top two plots ((a) and (b)) and slide the functions up and down. Describe what happens. Pay particular attention to fixed points. Something different happens to the lower two plots ((c) and (d)) when you do this? Describe the difference.

Exploration 48. Pretend that you could reach into the lower two plots and twist them around $x = 0$ in one direction or the other (by this I mean you can push one side up and one side down but the graph is pinned to $x = 0$). Again, describe what happens. Pay particular attention to fixed points. Something different happens to the upper two plots when you do this? Describe the difference.

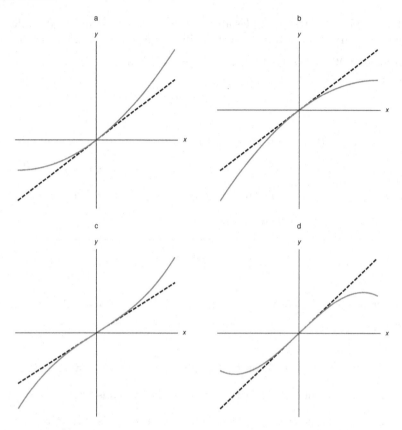

FIGURE 3.2
Plots of 4 functions that have neutral fixed points.

It will be helpful if we give some names to the dynamics that you observed in the explorations above. The easiest is captured in the bottom two figures. We call a fixed point **weakly attracting** if it is neutral fixed point and there exists an interval U containing the fixed point such that initial conditions in U converge to the fixed point under iteration. **Weakly repelling** is defined similarly. There is no standard name for the dynamics in the top two figures. I'm tempted to combine the words "repelling" and "attracting" and call them *retracting* points but that has a different meaning. So let's use the less clever term of **directionally attracting.**

Conjecture

Keeping in mind the examples above, let's put together all of the conditions needed for a fixed point to be directionally attracting or weakly attracting.

Conjecture 49. Let's start with directionally attracting (graphs (a) and (b) of figure 3.2).

If

- _____ (the condition that implies that $x = 0$ is a fixed point)

- _____ (the condition that implies that $x = 0$ is neutral) and

- _____ (the condition that implies that $x = 0$ is directional)

then $x = 0$ is a directionally attracting fixed point.

How does the sign of $f''(0)$ determine which side is attracting and which side is repelling?

Conjecture 50. Now let's turn to the graphs (c) and (d) in figure 3.2.

If

- _____ (the condition that implies that $x = 0$ is a fixed point)

- _____ (the condition that implies that $x = 0$ is neutral) and

- _____ (the condition that implies that $x = 0$ is attracting)

then $x = 0$ is a weakly attracting fixed point.

- If _____ then $x = 0$ is a weakly repelling fixed point.

Apply

When studying the dynamics of a family of functions that depend on a single parameter, it is often important to find both the variable and parameter values where that families of functions $f_a(x)$ have neutral fixed points. There is a system of two equations that need to be solved simultaneously to do this:

$$
\begin{aligned}
f_a(x) &= x \\
|f_a'(x)| &= 1.
\end{aligned}
$$

Let's return to our investigation of the logistic equation for exercises 51 and 52.

Application 51. [IBLdynamics.com] For what value of the parameter a is the fixed point $x = 0$ neutral? This value of a is called a **bifurcation value**. Describe the stability of this fixed point for values of a on either side of the bifurcation value. The logistic iteration tool on the website can be used to help you with this but the calculations should be done by hand as well.

Application 52. [IBLdynamics.com] For what value of the parameter a is the non-zero fixed point $x = (a - 1)/a$ neutral? Describe the stability of this fixed point for values of a on either side of the bifurcation value. Use the logistic iteration tool on the website to help you with this, but the calculations should be done by hand as well.

Prove

We won't be proving either of the following theorems. However, we think that it is important to list them for future reference. Note that in conjecture 49, we assumed for simplicity that the fixed point was at $x = 0$. In what follows, the fixed point is $x = \tilde{x}$.

Theorem 3.4 *If*

- $f(\tilde{x}) = \tilde{x}$,

- $f'(\tilde{x}) = 1$, *and*

- $f''(\tilde{x}) \neq 0$

then $x = \tilde{x}$ *is a directionally attracting fixed point.*

Theorem 3.5 *If*

- $f(\tilde{x}) = \tilde{x}$,

- $f'(\tilde{x}) = 1$,

- $f''(\tilde{x}) = 0$, *and*

- $f'''(\tilde{x}) < 0$

then $x = \tilde{x}$ *is a weakly attracting fixed point. If* $f'''(\tilde{x}) > 0$, *then the fixed point is weakly repelling.*

Proof 53. Explain why the last condition of Theorem 3.5 ($f'''(\tilde{x}) < 0$) is necessary.

We have focused on neutral fixed points where $f'(\tilde{x}) = 1$ in the examples above but neutral fixed points also occur when $f'(\tilde{x}) = -1$. This is why we needed to solve $|f_a'(x)| = 1$ in the *Apply* section above.

3.2   Periodic Points

Periodic points and orbits were introduced in chapter 1. In this section, we discuss them further. Our primary emphasis will be determining the stability of a periodic point. Recall that in definition 1.4 we said that a point x_0 is a **period p point** of a dynamical system $x_{n+1} = f(x_n)$ if

$$f^p(x_0) = x_0. \tag{3.7}$$

If p is the smallest integer for this equality to hold, then x_0 is a **prime period p point**. (Note that the adjective prime here does not modify the number p and thus we are not saying that the number p is prime! A function might have a prime period 4 orbit for example.) Additionally, if x_0 is a period p point, then the associated **periodic orbit** is

$$\{x_0, x_1, \ldots, x_{p-1}, x_0, \ldots\}$$

where $f(x_k) = x_{k+1}$ for $k = 0, 1, \ldots, p-2$ and $f(x_{p-1}) = x_0$.

In addition to the stability question, there is a second important question about periodic orbits that we will begin to explore here. If a given dynamical system has a period p point, what other periodic points must also exist? The answer to this question will ultimately be answered in chapter 10, but we will be able to give a partial answer in this section.

3.2.1   Stability of Periodic Points

Explore

The following questions explore some of the basic properties of periodic orbits.

Exploration 54. Suppose that x_0 is a prime period p point. What are all of its non-prime periods? Explain.

Exploration 55. Suppose that x_0 is a period p point (not necessarily prime). What periods *must* it also be? What periods *could* it also be?

Exploration 56. Suppose that you know that a dynamical system f has a prime period 2 orbit and you are given a graph of $y = f^2(x)$. How could you locate those points using the graph? How do you make sure that they are the prime period 2 points? How does this concept generalize to prime period p points?

Exploration 57. IBLdynamics.com The tool on the website generates a pair of stair-step diagrams for the logistic equation with an initial condition of $x_0 = .5$. Both $y = f_a(x)$ and $y = f_a^n(x)$ are shown.

For each of the a-values given below, the logistic equation has an attracting periodic orbit. Determine its period. How many periodic points are there of this period? How many periodic orbits are there of this period?

---

a) $a = 1.75$ b) $a = 3.25$ c) $a = 3.46$ d) $a = 3.63$ e) $a = 3.84$

Now let's turn our attention to determining stability of periodic orbits. If x_0 is a prime period p point of a function f, then the definition of periodicity implies that x_0 is a fixed point of $f^p(x)$. This is a really important fact! Keep it in mind as you complete the exercise below.

Exploration 58. IBLdynamics.com Use the parameter values from the previous exercise to answer these questions. Writing down the period of each attracting periodic orbit you found, use the tool on the website to look at the graph of $f_a^n(x)$ for the correct value of a and period n. What can you say about the derivative of the right graph at the attracting periodic points? What about the repelling ones?

Conjecture

Conjecture 59. You know that if a differentiable function f has a fixed point x_0 and $|f'(x_0)| < 1$, then x_0 is attracting (and if this value is greater than 1, it is repelling). Now, suppose that f has a prime period p point x_0. Complete the following theorem that gives conditions on when a periodic point x_0 is attracting or repelling?

> If f is differentiable, $f^p(x_0) = x_0$, and _____ then x_0 is an attracting period p point. Similarly, if _____, then x_0 is a repelling period p point.

Apply

Application 60. Let $f(x) = x^2 - 2$. What is $f^2(x)$ and what degree polynomial is it? What is $f^3(x)$ and what degree polynomial is it? In general, what is the degree of $f^n(x)$? What does this tell you about the task of computing periodic points?

Application 61. The function $f(x) = x^2 - 1$ has both fixed points and period 2 points. Find them and determine their stability.

Prove

Theorem 3.6 *If f is differentiable, $f^p(x_0) = x_0$, and*

$$\left|(f^p)'(x_0)\right| < 1,$$

then x_0 is an attracting period p point. Similarly, if

$$\left|(f^p)'(x_0)\right| > 1,$$

then x_0 is a repelling period p point.

The proof of this is very simple. Simply apply the attracting fixed point theorem to the function f^p. However, we will work through a slight restatement of this theorem.

Proof 62. Suppose that f is differentiable and $f(x_0) = x_1$. Show that

$$\left(f^2\right)'(x_0) = f'(x_0) f'(x_1).$$

Proof 63. Suppose that f is differentiable and $\{x_0, x_1, \ldots, x_{p-1}\}$ is a prime period p orbit. Show that

$$(f^p)'(x_0) = f'(x_0) f'(x_1) \ldots f'(x_{p-1}).$$

Conclude that if the product of the derivatives evaluated on the periodic orbit is less than 1 in absolute value, then the orbit is attracting.

3.2.2   New Periodic Orbits from Old

Back in chapter 1, we discussed that one of our overarching goals is to completely describe the dynamics of a given dynamical system. One of the things we want to include in this description is the types of periodic orbits that do or do not arise in a given dynamical system. It turns out that for continuous functions of a real variable, we can deduce the existence of some periodic orbits simply by knowing the existence of a periodic orbit of some other period. We aren't quite ready to answer this question completely, but the tool we've been playing with and the fixed point theorems from above can be used to start answering this question.

Explore

Exploration 64. $\boxed{\texttt{IBLdynamics.com}}$ In exploration 57, you determined attracting periodic orbits for the logistic map. This is summarized in the table below.

a-value	period
1.75	1
3.25	2
3.46	4
3.63	6
3.84	3

TABLE 3.1

For each of these parameter values, complete the following using the periodic orbit tool on the website.

1. Determine whether there are prime period n orbits for $n = 1, \ldots, 10$. As n gets bigger, it may be difficult to see (and count) all intersections of the functions graph with the diagonal. Do the best you can. For each parameter, make a table that summarizes what other *prime periodic orbits* exist for that particular value.

2. Which of these parameter values gave you the most periodic orbits? What was the period of the attracting orbit for that parameter value?

Conjecture

Conjecture 65. Based on the explorations above, do you believe that if f has a periodic point, then f must also have a fixed point? Why or why not? Do you believe that if f has a fixed point then it must have a periodic point of prime period greater than 1? Why or why not?

Prove

Now we turn to proving some of the things that you observed in the explorations above. Your main tool for these proofs is Theorem 4.1 and using the fact that $f^k(f^m(x)) = f^{k+m}(x)$. You first encountered this fact in exploration 6 of chapter 1.

Proof 66. Prove the following. If f is continuous and f has a prime period 2 orbit, then f has a fixed point. (You've already done this! In what problem?)

There are two nice corollaries to the theorem you just proved.

Proof 67. Prove the following. If f is continuous and has a prime period $2n$ orbit, then f has a period n orbit. **Hint:** Let $G(x) = f^n(x)$. What is $f^{2n}(x)$ in terms of G.

Proof 68. Prove the following. If f is continuous and has a prime period 2^n orbit, then f has periodic orbits of period 2^k for all $k \leq n$. **Hint:** How can you use proof 67?

4

Analysis of Fixed Points

4.1 Fixed Point Existence Theorems

There is a rich mathematical history of fixed point theorems. It should be clear from an application perspective why this is so: a fixed point corresponds to an equilibrium state of the system being modeled and a natural question to ask about any natural phenomenon is "does the system reach equilibrium and if so, what is it?"

Many of these fixed point theorems apply to a broad array of underlying state spaces. For example, there are fixed point theorems for functions whose domain and range is the real numbers; for functions whose domain and range is a closed ball; and on and on. Probably, the two most celebrated fixed point theorems are the *Brouwer fixed point theorem* and the *contraction mapping theorem*.

As you have learned in the exercises of chapter 3, there are basically two fixed point questions to be answered about a given dynamical system. We first need to determine whether the dynamical system has at least one fixed point. This is naturally called an **existence question.** Once we determine that a fixed point does exist, then we turn to questions of stability. Is it attracting or repelling, and if it is attracting, what values converge to the fixed point (this is called the **basin of attraction**)?

The existence question is generally the easiest to answer. However, it is often more difficult to determine either the number of fixed points or their stability. For example, the Brouwer fixed point theorem can be used to determine the existence of fixed points for a surprisingly broad array of dynamical systems. However, it says nothing about whether there is one, two, or two hundred fixed points. And if it can't do that, it certainly can't tell us anything about their stability!

In this section, you will explore two different fixed point theorems for functions on $\mathbb{R}$. The first one is like the Brouwer fixed point theorem in that it can only be used to determine whether a fixed point exists. The second one, like the contraction mapping theorem, tells us there is exactly one fixed point and it is attracting. This additional information does, however, come at a cost.

The Brouwer Fixed Point Theorem

The Brouwer fixed point theorem states that every continuous function from a closed ball of $\mathbb{R}^n$ to itself has at least one fixed point. On the real line, a closed ball is more commonly called a closed interval and this is what you will be proving below. In $\mathbb{R}^2$, a closed ball is a closed disk. One aspect of this theorem that makes it powerful is that it that it is true in $\mathbb{R}^n$ and not just $\mathbb{R}$.

The wonderful book *Differential Topology* [4] provides a very nice proof of this theorem and demonstrates it application in some very interesting examples.

Explore

Both of the theorems in this section concern functions that map a closed interval $[a, b]$ into itself. In other words, the output interval of the function is a subset of the input interval to the function. Here is a simple, yet surprisingly useful, way of visualizing such function that begins with drawing a rectangle using the directions below.

- Sketch a set of axes and include on it the diagonal line $y = x$. On the x-axis, mark the two endpoints of the interval $[a, b]$.

- Use the line $y = x$ to mark the points (a, a) and (b, b). Also, mark the values a and b on the y-axis.

- Use the points (a, a) and (b, b) to draw a rectangle. When you are finished, you should have picture similar to figure 4.1.

Exploration 1. Suppose that f maps the interval $[a, b]$ into the interval $[a, b]$. What is the relationship between the graph of $f(x)$ and the rectangle that you constructed?

Exploration 2. Sketch a function $f(x)$ that maps the interval $[a, b]$ into the interval $[a, b]$ that has *exactly one fixed point*. Sketch a function $f(x)$ that maps the interval $[a, b]$ into the interval $[a, b]$ that has *multiple fixed points*.

Exploration 3. Sketch a function $f(x)$ that maps the interval $[a, b]$ into the interval $[a, b]$ that has *no fixed points*. What property must the function f lack to make this possible?

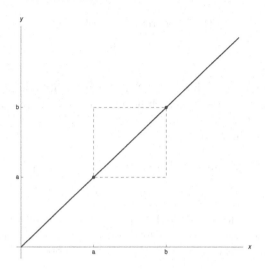

FIGURE 4.1
The line $y = x$ and the rectangle created using the two corner points (a, a) and (b, b).

Conjecture

Conjecture 4. Based on these exercises, what is the missing hypothesis and conclusion in the following theorem?

> If $f([a, b]) \subset [a, b]$ and f is _____ , then there exists a point $c \in [a, b]$ such that _____ .

Apply

The following exercises explore some of the hypotheses and conclusions of conjecture 4 (which is given in Theorem 4.1 below). You are given functions, asked to determine whether the theorem does or does not apply, and then to determine whether there is or is not a fixed point. Think carefully about each problem because together they illustrate the strengths and limitations of this theorem.

Application 5. Verify that Theorem 4.1 applies to

$$f_k(x) = \frac{1 + \sin{(kx)}}{2}$$

on the interval $[0,1]$ for each $k = 1, 2, 3, \ldots$. The theorem implies that each f_k has a fixed point in $[0,1]$. Does each have a unique fixed point in the interval $[0,1]$? Plot several of these and discuss what you see and how that does or does not relate to the theorem.

Application 6. Explain why Theorem 4.1 *does not* apply to

$$g_k(x) = 1 + \sin{(kx)}$$

on the interval $[0,1]$ for each $k = 1, 2, 3, \ldots$? Does this mean that there is no fixed point in the interval $[0,1]$? Plot several of these and discuss what you see and how that does or does not relate to the theorem.

Application 7. Explain why Theorem 4.1 *does not* apply to

$$h(x) = \begin{cases} x + .5 & \text{if } 0 \le x < .5 \\ x - .4 & \text{if } .5 \le x \le 1 \end{cases}.$$

Does this mean that there is no fixed point in the interval $[0,1]$? Plot this function and discuss what you see and how that does or does not relate to the theorem.

Application 8. Explain why Theorem 4.1 *does not* apply to

$$q(x) = \begin{cases} .25x + .25 & \text{if } 0 \le x < .75 \\ x - .75 & \text{if } .75 \le x \le 1 \end{cases}.$$

Does this mean that there is no fixed point in the interval $[0,1]$? Plot this function and discuss what you see and how that does or does not relate to the theorem.

Prove

Theorem 4.1 *If $f([a,b]) \subset [a,b]$ and f is continuous, then there exists a point $c \in [a,b]$ such that $f(c) = c$.*

After drawing the pictures, the proof of this statement probably seems so obvious that it hardly seems worth doing. However, there is a bit of work to do.

Proof 9. Find your old calculus book and look up the Intermediate Value Theorem. Use it to prove this theorem.

Explore

Theorem 4.1 is an existence theorem in that it provides a condition that guarantees that at least one fixed point exists in a given interval. It does not tell us how many nor does it tell us anything about the stability of these fixed points. In the next set of explorations, we will try and take this idea a bit further in an attempt to address these limitations.

Exploration 10. Make additional copies of figure 4.1 (two or three will suffice). Sketch several functions that map the interval $[a, b]$ into the interval $[a, b]$. They should be increasing and "less steep" than the line $y = x$.

1. How many fixed points do these functions have?

2. Use graphical analysis to determine the stability of the fixed points in this case. What initial conditions (if any) in the interval $[a, b]$ converge to the fixed points?

3. What is the formal mathematical way of saying that a function f is everywhere increasing and "less steep" than the line $y = x$?

Exploration 11. How many fixed points must a decreasing function that maps the interval $[a, b]$ into the interval $[a, b]$ have? Repeat the last two items of exploration 10 in this case.

Conjecture

Conjecture 12. Thinking about the pictures you drew above, complete the following statement.

If $f([a, b]) \subset [a, b]$ and _____ , then f has a _____ fixed point in $[a, b]$ that is _____.

Apply

Application 13. Suppose that a point $x = a$ is a prime period 2 point of f with $f(a) = b$, $f(b) = a$ and $a < b$. Prove that if f is continuous and has a prime period 2 orbit, then f has a fixed point as well.

Prove

Below is a slightly more formal version of the theorem of conjecture 12.

Theorem 4.2 *If f is differentiable with a continuous derivative, $f([a,b]) \subset [a,b]$, and $|f'(x)| < 1$ for all $x \in [a,b]$, then f has a unique fixed point $\tilde{x} \in [a,b]$. Moreover, if $x_0 \in [a,b]$, then $f^n(x_0) \to \tilde{x}$ as $n \to \infty$.*

Below are the exercises that lead you to the proof of this theorem. But before completing those steps, it is instructive to think about a "non-proof" that seems easy, but is unfortunately incorrect.

Proof 14. The following "proof" is wrong for at least two reasons! What are they and why?

Since $f([a,b]) \subset [a,b]$, we know by the Theorem 4.1 that f has a fixed point $\tilde{x}$ in the interval $[a,b]$. By assumption, $|f'(\tilde{x})| < 1$. This implies that the fixed point is attracting. Thus, if $x_0 \in [a,b]$, then

$$\lim_{n \to \infty} f^n(x_0) = \tilde{x}.$$

Now, let's do it right in proofs 15 through 18.

Proof 15. Adapt the proof of Theorem 4.1 and use the given derivative restriction to prove that f has a unique fixed point $\tilde{x} \in [a,b]$.

Proof 16. Look up the Mean Value Theorem in your calculus book. Explain why that theorem and the derivative condition given here implies that there exists $\lambda \in (0,1)$ such that

$$|f(x) - f(y)| \leq \lambda|x - y|$$

for all x and y in the interval $[a,b]$.

Proof 17. Now fix an initial condition $x_0 \in [a,b]$ that is not the fixed point. Show that the sequence of iterates $\{f^n(x_0)\}$ is a Cauchy sequence (this requires using the Mean Value Theorem inequality from proof 16 and the geometric sum formula). Explain why this sequence must converge to a point $x' \in [a,b]$.

Proof 18. Complete the proof by using the continuity of f to show that x' is a fixed point of f.

The Contraction Mapping Theorem

Theorem 4.2 is a special case of the **contraction mapping theorem**; a fixed point theorem that applies to functions not just on $\mathbb{R}$ but on any *complete metric space*. One application of the contraction mapping theorem that you may have encountered elsewhere is in the proof of the existence and uniqueness theorem in a differential equations course.

A metric space is a topological space where there is a way of measuring the distance between any two points in the space. Let's call this distance function (or metric) d so that $d(x, y)$ is the distance between x and y. As you read in chapter 2, a metric space is complete if every Cauchy sequence in the space converges to a point in the space.

A function f is a **contraction** if there exists a $\lambda \in (0, 1)$ such that

$$d(f(x), f(y)) \leq \lambda d(x, y)$$

for all points x and y in our metric space. The contraction mapping theorem says that if f is a contraction on a complete metric space, then f has a unique fixed point. The proof of this much more general theorem is almost identical to the proof that you completed in the exercises above.

4.2 The Inverse and Implicit Function Theorems

The implicit function theorem is the key tool in understanding basic bifurcation theory, the topic of chapter 5. In this section, you will complete a sequence of explorations aimed at understanding this theorem. We will not prove this theorem here, but note that it is often covered in an advanced calculus or multivariable analysis course. Because you encountered it previously in both calculus and linear algebra, we begin with an exploration of the an equivalent, but more familiar theorem: the inverse function theorem.

4.2.1 The Inverse Function Theorems

The inverse function theorem is presented in calculus about the time you learn the derivative formulas for $\ln(x), \arctan(x)$, and other inverse functions. You may not realize it, but you also encountered this theorem in linear algebra (although it is usually not named this). The goal of this section is two-fold. First, we hope to re-familiarize you with invertibility and think more carefully about this concept. Second, we want to connect the calculus and linear algebra versions of this theorem. This should set the stage for understanding the implicit function theorem. So, let's start with the calculus version.

Explore

Exploration 19. When is a function $f : \mathbb{R} \to \mathbb{R}$ invertible?

Exploration 20. Describe some of the relationships between the original function f and its inverse f^{-1}.

1. What happens when you compose f and f^{-1} in either order?
2. What are their domains and ranges and how are they related to each other?
3. How are their graphs related to each other?

Exploration 21. Let's look at a concrete example. Suppose that $f(x) = 5x + 4$. What is $f^{-1}(y)$? Describe a general process for computing $f^{-1}(y)$ given a function $f(x)$. Is this process always "doable"?

Exploration 22. Now, graph the function $f(x) = x^2$. Is it invertible given your answer to exploration 19?

Exploration 23. What do we mean when we write "if $f(x) = x^2$ then $f^{-1}(x) = \sqrt{x}$"? What unsaid assumption did we make about the domain of f?

Exploration 24. If $f(x)$ is differentiable and invertible, what is a formula for $(f^{-1})'(y)$? If you don't remember this from calculus, use the chain rule to differentiate

$$f\left(f^{-1}(y)\right) = y$$

and solve for $(f^{-1})'(y)$. Remember that since $f(x) = y$ we have $f^{-1}(y) = x$.

Now, let's switch gears and think about the idea of invertibility from linear algebra.

Exploration 25. If A is an $n \times n$ matrix, when is A invertible? There are quite a few equivalent properties so you should list as many as you can remember.

Exploration 26. If $L(x) = Ax$ where $x \in \mathbb{R}^n$, when is the function L invertible? How does this differ, if at all, from your answer above?

Finally, let's try and relate these two ideas with an example. Consider the function $F : \mathbb{R}^2 \to \mathbb{R}^2$ defined by

$$F(x, y) = (x^2 + xy + y^2, 3x - y).$$

Exploration 27. The derivative DF of F is a 2×2 matrix of partial derivatives with the first row being the partial derivatives of the first function and the second row being the partial derivatives of the second function. Compute DF.

Exploration 28. Compute $DF(1,1)$ and $DF(0,0)$. One of these two matrices is invertible and one is not. Identify which is which.

Exploration 29. What do you think these calculations imply about the invertibility of the original function F?

Prove

Unlike other *Prove* sections, you will not be lead through a proof of this theorem. Instead, our goal is to simply present to you the theorem and to take some time to explain it and relate it to the work you did in the *Explore* section. We'll begin with the version of the inverse function theorem that you learned in calculus that applies to functions of 1 variable.

Theorem 4.3 (Inverse Function Theorem (single variable)) *Suppose that $f : \mathbb{R} \to \mathbb{R}$ is differentiable, $f(x_0) = y_0$, and $f'(x_0) \neq 0$. Then there exists open intervals $U \ni x_0$ and $V \ni y_0$ and a differentiable function $f^{-1} : V \to U$ such that*

- $f^{-1}(y_0) = x_0$,

- $f^{-1}(f(x)) = x$ *for all $x \in U$ and $f(f^{-1}(y)) = y$ for all $y \in V$, and*

- $\left(f^{-1}\right)'(y_0) = \dfrac{1}{f'(x_0)}$.

This version of the inverse function theorem says that near a non-critical point (sometimes called a *regular point*) there is an inverse function to f. In other words, there is a unique solution to $y = f(x)$ in that neighborhood. It does **not** say that you can compute the inverse function. We call this an *existence result* because it says such a thing exists, but gives us no method of constructing it. Nor does it say that this solution works everywhere; only in some possibly small interval about the given point. We call this a *local result* simply because it applies only near a given regular point. This is why this theorem and those like it, are called *local existence theorems*.

Theorem 4.3 generalizes to functions $F : \mathbb{R}^n \to \mathbb{R}^n$. In this version, we explicitly see the connection between the invertibility of the function F and the invertibility of the derivative matrix DF.

Theorem 4.4 (Inverse Function Theorem (multivariable)) *Suppose that $F : \mathbb{R}^n \to \mathbb{R}^n$ is differentiable, $F(x_0) = y_0$, and $DF(x_0)$ is an invertible matrix. Then there exists open sets $U \ni x_0$ and $V \ni y_0$ and a differentiable function $F^{-1} : V \to U$ such that*

- $F^{-1}(y_0) = x_0$,

- $F^{-1}(F(x)) = x$ *for all* $x \in U$ *and* $F(F^{-1}(y)) = y$ *for all* $y \in V$, *and*

- $DF^{-1}(y_0) = (DF(x_0))^{-1}$.

Like the first version of this theorem, this is a local existence theorem. It says that if the Jacobian matrix DF of F is invertible *at the point* x_0, then the function F is locally invertible. In exploration 25, you were asked to list a number of conditions that are equivalent to a matrix being invertible. One of those conditions is that the determinant should be non-zero. So, this theorem could have read "if $\det(DF(x_0)) \neq 0$ then ... ".

This is helpful in understanding why the second theorem is a generalization of the first. If $n = 1$, then DF is a 1×1 matrix, which is simply a number, and the determinant of a number is just that number. We know that numbers are invertible (i.e., have a reciprocal) as long as they are not equal to 0, which is exactly the condition given here.

4.2.2   The Implicit Function Theorem

Explore

This theorem is probably completely new to you although the underlying algebra is something that you've known for a while. The goal of these exercises is to help you see the connection between the algebra that you know well and this new theorem, which might sound quite intimidating.

For the following exercises, consider the function $G : \mathbb{R}^2 \to \mathbb{R}$ defined by

$$G(x, y) = x^2 + y^2 - 1.$$

Exploration 30. What is $\{(x, y) \mid G(x, y) = 0\}$? Describe this set both in words and in a picture.

Exploration 31. Can you solve $G(x, y) = 0$ for y?

Exploration 32. Can you use this solution to define a function? How?

Exploration 33. Does your answer work for the point $(\sqrt{2}/2, \sqrt{2}/2)$? What about the point $(\sqrt{2}/2, -\sqrt{2}/2)$?

Exploration 34. Use the function G to compute dy/dx using implicit differentiation. At what values of (x, y) is dy/dx undefined? Identify these points on your picture from exploration 30.

Exploration 35. Can you find a single solution to $G(x, y) = 0$ that works for **all** (x, y) in a neighborhood of the point $(1, 0)$? If so, what is it and what is the neighborhood? If not, what went wrong?

Exploration 36. Can you find a single solution to $G(x, y) = 0$ that works for **all** (x, y) in a neighborhood of the point $(\sqrt{2}/2, \sqrt{2}/2)$? If so, what is it and what is the neighborhood? If not, what went wrong?

Exploration 37. Compute the partial derivatives of G and evaluate them at both of the points above. What happened in the case where you could not solve for y?

Exploration 38. Name the function that you found in problem 35 or 36 above $\gamma(x)$. What is $G(x, \gamma(x))$?

Prove

The inverse function theorem (4.3) tells us when we can solve $y = F(x)$ for x. The implicit function theorem is very similar. It tells us when we can solve $G(\lambda, x) = 0$ for x. We will usually be interested in cases where λ is a real number, but the theorem is the same if λ is a list of real numbers. Thus, we will state the theorem in that setting.

Theorem 4.5 (Implicit Function Theorem) *Suppose that* $G : \mathbb{R}^k \times \mathbb{R} \to \mathbb{R}$; $(\lambda, x) \mapsto G(\lambda, x)$, *is a* C^1 *function satisfying*

- $G(\lambda_0, x_0) = 0$, *and*

- $\dfrac{\partial G}{\partial x}(\lambda_0, x_0) \neq 0.$

Then there exists $\delta > 0$ *and a* C^1 *function*

$$\psi : \left\{ \lambda \in \mathbb{R}^k \mid |\lambda - \lambda_0| < \delta \right\} \to \mathbb{R}$$

such that $\psi(\lambda_0) = x_0$ *and* $G(\lambda, \psi(\lambda)) = 0$ *for all* λ *with* $|\lambda - \lambda_0| < \delta.$

Again, note that the implicit function theorem is a local existence theorem. It does not tell us how to solve for x, simply that a solution does exist in a neighborhood of a given regular point. The function ψ of the theorem is that solution.

There are a few more things to note about the implicit function theorem. First, this is actually a special case of the theorem. There is a version of it that allows x to be of dimension greater than one similar to the second version of the inverse function theorem. In this more general version, the ability to solve for x again depends on a linear algebra condition of a matrix of partial derivatives. We will not be needing this version here. Second, the implicit function theorem does give us a formula for the derivative of the function ψ. We have chosen to omit it here because we will not need it to prove the upcoming bifurcation theorems.

Finally, and maybe this is not a surprise, the inverse and implicit function theorems are equivalent. That means you can prove one of them by assuming that the other one is true. This is actually not that difficult and I encourage you to try and prove that the single variable inverse function theorem (4.3) is equivalent to the implicit function theorem (4.5) with $\lambda \in \mathbb{R}$.

4.3   Hyperbolic Periodic Points

In dynamical systems, we are often interested in describing more than the dynamics of a single system. Often we want to describe the range of dynamical behaviors that occurs in a family of equations that depend on parameters. A typical question to ask, from a mathematical modeling perspective, is "do the dynamics observed for one specific parameter value persist for other nearby parameter values?" If not, then the model is probably not a very good description of the observed phenomenon.

Sometimes, this persistence of dynamics with respect to changes to a parameter is referred to as **structural stability**. Here is a simple example that you have already explored. You know that in the logistic family that $x = 0$ is an attracting fixed point when $0 < a < 1$. Thus for a parameter value in this range, the logistic equation is structurally stable. A small change to the parameter does not significantly alter the dynamics. Similarly, when $1 < a < 3$, the system is structurally stable. For a value of a in this range, a small change to the parameter moves the fixed point slightly, but does not change the overall dynamics.

The following definition will help us quantify the idea of structural stability.

Definition 4.1 *A fixed point x_0 of a differentiable dynamical system $x_{n+1} = f(x_n)$ is **hyperbolic** if $|f'(x_0)| \neq 1$. If $|f'(x_0)| = 1$, then the fixed point is* **non-hyperbolic**. *A period p point x_0 is hyperbolic if $|(f^p)'(x_0)| \neq 1$.*

Explore

Recall that a fixed point of a function $f(x)$ is simply the intersection of the graph of $y = f(x)$ and the line $y = x$. If we have a family of functions f_a, then the fixed points are again the intersections of the graph of $y = f_a(x)$

and the line $y = x$ but the exact location of these fixed points may change depending on the value of parameter a. The following exercises explore what happens to fixed points as a parameter changes. Use the apps on the website to answer the questions below.

Exploration 39. | IBLdynamics.com | Before you use any of the website tools, look at the graph on the site. Use it to estimate the value of the fixed point of this function and the value of the derivative of the function at the fixed point.

1. According to definition 4.1, is the fixed point of this function hyperbolic or non-hyperbolic?

2. As you interact with the graph, does the fixed point move? Do any new fixed points appear? Is this new fixed point(s) hyperbolic or non-hyperbolic?

3. Is this system structurally stable?

4. Can you manipulate the graph in such a way that there are multiple fixed points? Do you have to manipulate the graph a little or a lot?

Let's repeat this exercise but with a slightly different starting figure.

Exploration 40. | IBLdynamics.com | Before you use any of the website tools, look at the graph on the site. Use it to estimate the value of the fixed point of this function and the value of the derivative of the function at the fixed point.

1. According to definition 4.1, is the fixed point of this function hyperbolic or non-hyperbolic?

2. As you interact with the graph, does the fixed point move? Do any new fixed points appear? Is this new fixed point(s) hyperbolic or non-hyperbolic?

3. Is this system structurally stable?

4. Can you manipulate the graph in such a way that there are multiple fixed points? Do you have to manipulate the graph a little or a lot?

The exercises above should have given you a good intuitive understanding of the relationship between hyperbolic fixed points and structural stability. Wiggling the graph a little bit (that is changing the parameter slightly) does not cause a new fixed point to emerge or the existing one to vanish. However, we need to be more precise than this and that degree of precision is the main difficulty in discovering the statement of this theorem. One needs to precisely state the hypotheses and conclusions using formal mathematical language and notation. The following exercises are designed to walk you through the process of expressing this concept precisely in a step-by-step fashion. In the final exercise, you are prompted to put all of this together in a formal statement of the Hyperbolic Fixed Point Theorem.

Exploration 41. Express the statement

> when $c = c_0$ the family of functions $F_c(x)$ has a hyperbolic fixed point $x = x_0$

in formal mathematical terms and notation.

Exploration 42. Now suppose that along with the conditions above, $F_c(x)$ is continuous and differentiable with respect to both c and x. Informally, what can you say about fixed points of $F_c(x)$ for $c \neq c_0$ but nearby? What can you say about the derivative at the fixed point for $c \neq c_0$ but nearby?

Exploration 43. How can we make the statement "for $c \neq c_0$ but nearby" formal using mathematical language and notation? (**Hint:** Remember, absolute value means distance!)

Exploration 44. How can we express the statement "there is a fixed point that depends on the value c" using mathematical notation?

Conjecture

Conjecture 45. Use your answers from above to write the **Hyperbolic Fixed Point Theorem**.

Suppose that x_0 is a hyperbolic fixed point of $F_{c_0}(x)$ (i.e., _____).
Then there exists a $\delta > 0$ and a C^1 function $\gamma : (c_0 - \delta, c_0 + \delta) \to \mathbb{R}$ such that

- _____ (conclusion about a fixed point), and

- _____ (conclusion about the derivative of the fixed point).

Apply

The logistic function of population growth can easily be modified to incorporate terms that take harvesting into account. In the case of proportional harvesting, the model equation becomes

$$x_{n+1} = ax(1 - x) - hx$$

and if the harvesting rate is constant, we get a second model

$$x_{n+1} = ax(1 - x) - h.$$

In both models, the parameter $h > 0$.

Exploration 46. Complete the following for both models.

1. Determine the fixed points. These will depend on both a and h.

2. You know that when $h = 0$, there is a non-zero, attracting, hyperbolic fixed point for $1 < a < 3$. What happens to the location of this fixed point as h increases from 0? What happens to the stability of this fixed point as h increases from 0?

3. Sketch the regions in the (a, h) plane where each of the fixed points is attracting. If you take a value on the boundary of one of these regions, is there still a fixed point? Is it hyperbolic or non-hyperbolic?

Exploration 47. Modify the definition of a hyperbolic fixed point to define a hyperbolic periodic point.

Prove

Theorem 4.6 (Hyperbolic Fixed Point Theorem) *Suppose that x_0 is a hyperbolic fixed point of $F_{c_0}(x)$ (i.e., $F_{c_0}(x_0) = x_0$ and $|F'_{c_0}(x_0)| \neq 1$). Then there exists a $\delta > 0$ and a C^1 function $\gamma : (c_0 - \delta, c_0 + \delta) \to \mathbb{R}$ such that*

- $F_c(\gamma(c)) = \gamma(c)$ *for all $c \in (c_0 - \delta, c_0 + \delta)$ (i.e., $\gamma(c)$ is a fixed point), and*

- $|F'_{c_0}(x_0)| \neq 1$ *(i.e., $\gamma(c)$ is a hyperbolic fixed point.)*

Before proceeding with the proof of this theorem, it will be helpful to discuss what exactly it says. First, the function $\gamma(c)$ is what we might call a "fixed point function." The input is the parameter c and the output is a fixed point of $F_c(x)$. Review the first bullet point to see why this is so. This is again a local existence theorem. It only tells us that this fixed point exists for parameter values near the original parameter value c_0. That is the role of the δ in the theorem. In the language of exploration 43, the interval $(c_0 - \delta, c_0 + \delta)$ formally defines what we mean by "for $c \neq c_0$ but nearby."

Now on with the proof.

Proof 48. Define a new function $G(c, x) = F_c(x) - x$. What is $G(c_0, x_0)$? What is

$$\left. \frac{\partial}{\partial x} G(c, x) \right|_{(c_0, x_0)}.$$

Proof 49. Use the Implicit Function Theorem to show that you can solve $G(c, x) = 0$ for x in terms of c near $c = c_0$. Call this function $\gamma(c)$.

Proof 50. Interpret this result in terms of the original family of functions F_c.

5

Bifurcations

5.1 What is a Bifurcation?

In the previous chapters, you were given some tools to analyze a given dynamical system. You know how to compute fixed points and determine whether they are attracting, repelling, or neither. You can do the same with periodic points although the algebra is generally much more tedious. In this section, our focus changes from considering a single dynamical system to considering families of dynamical systems that depend on parameters. You've been introduced to two different families that illustrate many of the fundamental concepts of bifurcation theory. Most of your work has focused on the logistic family

$$x_{n+1} = f_a(x_n) = ax_n(1 - x_n) \tag{5.1}$$

where the parameter is a. We indicate the function's dependence on a with a subscript to the function name. Less time has been spent studying the quadratic family

$$x_{n+1} = Q_c(x_n) = x_n^2 + c. \tag{5.2}$$

Here, c is the parameter and again denoted with a subscript.

When working with these families in the previous chapters, you were usually given a parameter value and then asked to find periodic points and determine their stability. Now, we will build on that work by asking these same questions; but now the answer will depend on the parameter. Let me give you a few example questions for the logistic family (though the same questions could be asked for the quadratic family or any other family of functions).

- For what values of a is the fixed point $x = 0$ attracting? How about repelling?

- There is non-zero fixed point as well. How does it depend on the parameter a? When is it attracting and when is it repelling?

- For what values of a are there period 2 orbits? How many? Attracting or repelling?

I think you get the basic idea here. We want to compute the values of the fixed points and periodic points in terms of the parameter. We also want to determine the stability of these points in terms of the parameter. Ultimately, our goal is to paint a picture (literally) that will describe the dynamics of f_a for every parameter value a. This picture is called a *bifurcation diagram* and you will begin to draw one in section 5.2.

But before we do that, we should first think about not how the parameter affects the dynamics, but instead how the parameter affects the graph of the function that determines the dynamics.

Explore

Consider the graphs of the family of quadratic functions defined in equation 5.2 when answering the following questions. There is a tool on the website, which may be helpful when answering these questions, but you should be able to answer all of them without using it by applying techniques from algebra and calculus.

Exploration 1. $\boxed{\texttt{IBLdynamics.com}}$ How does changing the parameter c affect the graph of $y = Q_c(x)$?

Exploration 2. $\boxed{\texttt{IBLdynamics.com}}$ What is the critical point of $Q_c(x)$ as a function of the parameter c? Is it a max or a min? What is the critical value (i.e., the y-value of the critical point)?

Exploration 3. $\boxed{\texttt{IBLdynamics.com}}$ What are the x and y intercepts as a function of c? For what values of c are there x-intercepts? What happens to the x-intercepts as c increases?

Exploration 4. $\boxed{\texttt{IBLdynamics.com}}$ We know that the intersections of the graph of $y = Q_c(x)$ and the line $y = x$ correspond to fixed points of the dynamical system. For what values of c are there fixed points? What are formulas for them in terms of c? Describe what happens to the fixed points as c *decreases*. In particular, how does the distance between them change, if at all?

Let's repeat these exercises, but this time with the graphs of the logistic family defined in equation 5.1. As in the last explorations, the tool on the website may be helpful, but you should be able to answer all of these questions without using it by applying techniques from algebra and calculus.

Exploration 5. $\boxed{\texttt{IBLdynamics.com}}$ How does changing the parameter a change the graph of $y = f_a(x)$?

Exploration 6. $\boxed{\texttt{IBLdynamics.com}}$ What is the critical point as a function of the parameter a? Is it a max or a min? What is the critical value (i.e., the y-value of the critical point)?

Exploration 7. $\boxed{\texttt{IBLdynamics.com}}$ What are the x and y intercepts of $f_a(x)$ in terms of a? For what values of a are there x-intercepts? What happens to the x-intercepts as a increases?

Exploration 8. $\boxed{\texttt{IBLdynamics.com}}$ For what values of a are there fixed points? What are formulas for them in terms of a? Describe what happens to them as a *increases*.

5.2 Introduction to Bifurcation Diagrams

A bifurcation diagram is a plot that summarizes how the dynamics of a system change as the parameter changes. Ideally, it should tell us about all periodic points, both attracting and repelling, and any other complicated behavior that may occur. In practice, this is not possible by hand calculation alone. Later, we will learn how to numerically generate bifurcation diagrams that completely describe all of the attracting dynamics (at least for the quadratic and logistic families and dynamical systems that are similar to those two families). Right now, the goal is simply to introduce the concept of bifurcation diagrams in the context of fixed points only.

Explore

The following exercises will lead you through the construction of a "fixed point only" bifurcation diagram for the logistic family with the parameter $a > 0$. You may want to refer to your answers from exercises 26 and 27 in subsection 3.1.2.

Exploration 9. Get a sheet of paper and two colored pens (say red and blue). The colors will be used to differentiate between attracting (red) and repelling (blue) fixed points.

1. On the sheet of paper, draw a set of axes with the horizontal axis labelled a (the parameter) and the vertical axis labelled x (the fixed point). Mark $a = 0, 1, 2, 3, 4$ on the horizontal axis.

2. You've already shown that one of the fixed points of the logistic function is $x = 0$. You also showed that this fixed point is attracting when $0 < a < 1$ and repelling when $a > 1$. Plot this curve (i.e., $x = 0$) on your axes using the blue pen for the attracting region and the green for the repelling.

3. You also showed that the other fixed point is given by $x = (a-1)/a$ and that it is attracting when $1 < a < 3$ and repelling for $a > 3$. Plot this curve on the same set of axes using the coloring scheme above.

Exploration 10. | IBLdynamics.com | Verify that you got the "fixed point only" bifurcation diagram for the logistic family that is on the website.

5.3   The Tangent Bifurcation

Probably, the most common bifurcation is the tangent or saddle-node bifurcation. Generally speaking, a tangent bifurcation occurs when an attracting fixed point and a repelling fixed point merge at some parameter value. Past this parameter value, there is no longer a fixed point. This description might seem a bit odd, but the following explorations should give you a good idea of how and why this happens. It is not as magical as it sounds.

Explore

The name "tangent bifurcation" gives a pretty good hint as to what happens in this kind of bifurcation. The first exercise illustrates the phenomenon.

Exploration 11. | IBLdynamics.com | There is a tool on the website to explore the tangent bifurcation. As you manipulate the graph on the site, pay particular attention to the appearance of fixed points. Adjust the value of the parameter c to answer the following questions.

1. Identify the value of c where the graph is tangent to the line $y = x$. How many fixed points are there at this value of c? What are their stabilities?

2. How many fixed points are there prior to this value of c? What are their stabilities?

3. How many fixed points are there after this value of c? What are their stabilities?

4. Make a rough sketch of the bifurcation diagram or "fixed point graph." Remember that the parameter c is the horizontal axis and the location of the fixed points is on the vertical axis. When you did this in exploration 9, you used color to indicate stability.

Exploration 12. ⊡IBLdynamics.com⊡ The two examples for this problem on the website are **not** tangent bifurcations. Use them as you did in exploration 11 to answer the following questions for each graph.

1. Identify the value of the parameter c where the graph is tangent to the line $y = x$.

2. How many fixed points are there prior to this value of c? How many fixed points are there after this value of c? What are their stabilities?

3. Draw the bifurcation diagram.

4. Explain what the difference is between these graphs and the graph of exploration 11. Can you use a single mathematical computation to explain why these are not tangent bifurcations?

Exploration 13. ⊡IBLdynamics.com⊡ The example for this problem on the website is again **not** a tangent bifurcation. Use it to answer the following questions:

1. Identify the value of the parameter c where the graph is tangent to the line $y = x$.

2. How many fixed points are there prior to this value of c? How many fixed points are there after this value of c? What are their stabilities?

3. Draw the bifurcation diagram.

4. Explain what the difference is between this graph and the graph of exploration 11. Can you use a single mathematical computation to explain why this is not a tangent bifurcation?

In exploration 11, you saw that a tangent bifurcation happens in a family of functions $F_c(x)$ when, at a bifurcation value $c = c_0$, there are no fixed points for values of c less than c_0, one fixed point at c_0, and two fixed points for values of c greater than c_0. Of course, this could also happen in the other direction; no fixed points for $c > c_0$, etc. Exercises 12 and 13 pointed to other necessary conditions for this bifurcation.

Exploration 14. The first condition is that

$$F_{c_0}(x_0) = x_0.$$

Explain this equation in plain English.

Exploration 15. The second condition is that at c_0, the graph of $y = F_{c_0}(x)$ must be tangent to the line $y = x$ at the fixed point $x = x_0$. Express this condition using formal mathematical notation.

Exploration 16. The third condition, illustrated in exploration 12, is that as the parameter changes through the bifurcation value, the graph of $y = F_c(x)$ must actually pass through the diagonal. In other words, you can't just "kiss" the diagonal. Express this condition using formal mathematical notation (**Hint:** It is again a derivative condition).

Exploration 17. Finally, the fourth condition implies that when the graph passes through the diagonal, *exactly* two fixed points appear. An example of this not occurring was exploration 13. Express this condition using formal mathematical notation (**Hint:** Reviewing subsection 3.1.4 might be helpful).

Take another look at the bifurcation diagram that you created in exploration 11. The curve you drew is not a function of the parameter c since it fails the vertical line test. But if we rotate this picture ninety degrees, it is a function. In other words, this "fixed point graph" can be viewed as function whose *inputs* are fixed point values (x) and whose *outputs* are the corresponding parameter value (c). This is different than what we did in the hyperbolic fixed point theorem. Let's call this function $p(x)$ so that at the bifurcation value we have $p(x_0) = c_0$.

Exploration 18. Look at the fixed point graph again. What can you say about $p'(x_0)$ and $p''(x_0)$?

Conjecture

Conjecture 19. Use your answers from above to write the **tangent bifurcation theorem**.

Consider the family of dynamical systems

$$x_{n+1} = F_c(x_n)$$

where $F_c : \mathbb{R} \to \mathbb{R}$ is a C^2 function in terms of both x and c. Additionally, assume that there exists a c_0 and x_0 such that

- $F_{c_0}(x_0) = x_0$,

- _____ (derivative condition from 15),

- _____ (derivative condition from 16), and

- _____ (derivative condition from 17).

Then there exists intervals $I \ni x_0$ and $J \ni c_0$ and a C^1 function $p : I \to J$ with $p(x_0) = c_0$ such that _____ (fixed point conclusion) for all $x \in I$. Moreover, _____ (derivative conclusions about p).

Apply

Application 20. The graph of the tangent bifurcation in exploration 11 is of the quadratic family of functions $Q_c(x) = x^2 + c$. Use the hypotheses of the tangent bifurcation theorem to find the exact value of c where the tangent bifurcation occurs. Draw the bifurcation diagram up to this point.

Application 21. Consider the family of dynamical system $x_{n+1} = e^x + c$. For what value of c is there a tangent bifurcation? Verify all of the hypotheses of the tangent bifurcation theorem.

Prove

Theorem 5.1 (Tangent Bifurcation Theorem) *Consider the family of dynamical systems*

$$x_{n+1} = F_c(x_n)$$

where $F_c : \mathbb{R} \to \mathbb{R}$ is a C^2 function in terms of both x and c. Additionally, assume that there exists a c_0 and x_0 such that

- $F_{c_0}(x_0) = x_0$,

- $\dfrac{\partial F_c}{\partial x}(c_0, x_0) = 1$,

- $\dfrac{\partial^2 F_c}{\partial x^2}(c_0, x_0) \neq 0$, *and*

- $\dfrac{\partial F_c}{\partial c}(c_0, x_0) \neq 0$.

Then there exists intervals $I \ni x_0$ and $J \ni c_0$ and a C^1 function $p : I \to J$ with $p(x_0) = c_0$ such that

$$F_{p(x)}(x) = x$$

for all $x \in I$. Moreover, $p'(x_0) = 0$ and $p''(x_0) \neq 0$.

We now proceed with the proof of the tangent bifurcation theorem. Let's simplify things a little bit by assuming that at the bifurcation value c_0, the fixed point is $x_0 = 0$. In other words, assume $F_{c_0}(0) = 0$. As we did in the proof of the hyperbolic fixed point theorem, let

$$G(c, x) = F_c(x) - x.$$

Proof 22. Show that $G(c_0, 0) = 0$.

Proof 23. Show that $\dfrac{\partial}{\partial c}G(c,x)\Big|_{(c_0,0)} \neq 0$.

Proof 24. Use the implicit function theorem to conclude that there exists a smooth function $p(x)$ such that $G(p(x),x) = 0$ near $x = 0$. Interpret this in terms of the original function F_c.

Note that you used the implicit function theorem here to solve for the parameter c in terms of the variable x. This is just the opposite of what was done in the proof of the hyperbolic fixed point theorem!

Proof 25. Differentiate $G(p(x),x) = 0$ with respect to x and solve for $p'(x)$. Use this to show that $p'(0) = 0$.

Proof 26. Differentiate the expression for $p'(x)$ again with respect to x. Use this to show that $p''(0) \neq 0$.

Proof 27. Explain what the graph of $c = p(x)$ represents in the bifurcation diagram. What do the derivative conditions above tell you about the shape of this graph?

5.4   The Period Doubling Bifurcation

To this point, we have explored two fundamental results about bifurcations. First, the hyperbolic fixed point theorem states that if the derivative at a fixed point is not equal to 1 in absolute value, then a bifurcation cannot occur at that particular parameter value. Second, the tangent bifurcation theorem describes what often happens when the derivative at the fixed point is equal to 1. So, what generally happens when the derivative at a fixed point is equal to -1?

If you go back and review the proof of the hyperbolic fixed point theorem (theorem 4.6), you will see that the proof still works if the derivative is equal to -1. Thus, a fixed point must persist for parameter values nearby. This is not surprising. The function illustrated in figure 5.1 has $f'(x_0) = -1$ at the fixed point x_0. Clearly, if we wiggle the graph around a little, the fixed point does not suddenly disappear. What does happen, however, is that the fixed point might change from being attracting to repelling or vice versa. But does anything else interesting happen when this occurs?

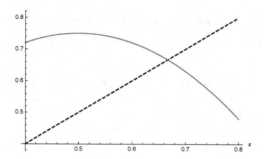

FIGURE 5.1

A plot of the graphs $y = f_3(x) = 3x(1 - x)$ and the line $y = x$. At the fixed point, $x_0 = 2/3$, $f_3'(x_0) = -1$. If we increase the parameter through $a = 3$, the fixed point remains though its stability changes from attracting to repelling.

Explore

Exploration 28. | IBLdynamics.com | Let's consider the logistic family of functions for values of $a \in [2.5, 3.3]$. The tool on the website generates stairstep diagrams in this parameter range, for initial condition $x_0 = 1/2$.

1. Show by direct calculation that when $a = 3$ the fixed point is $x_0 = 2/3$.

2. Show by direct calculation that $f_3'(2/3) = -1$.

3. Use the tool on the website to describe what happens to the dynamics as a increases from $a = 2.5$ to $a = 3.3$. In particular, what happens to the stability of the fixed point and why? Describe the dynamics near the fixed point for values of a less than 3 and for values of a greater than 3.

It is not at all clear from the exploration above where the period 2 orbit comes from. It seems to just emerge from the fixed point as the parameter passes through the bifurcation value at $a = 3$. If we want to understand something involving a period 2 orbit, it makes sense to look at the graph of $y = f_a^2(x)$. The following two explorations do that to illustrate what happens in a period-doubling bifurcation.

Exploration 29. $\boxed{\texttt{IBLdynamics.com}}$ The explorations on the website show two different plots of $y = f_a^2(x)$. Use these tools to answer the following questions.

1. Describe in general terms what happens to the non-zero fixed points of $f_a^2(x)$ as a increases through the bifurcation value of $a = 3$. How many fixed points are there when $a < 3$ and what are their stabilities? How many are there when $a > 3$ and what are their stabilities? The first tool will be most helpful for this exercise.

2. Recall that fixed points of $f_a^2(x)$ are period 2 points of $f_a(x)$. Are any of the fixed points of $f_a^2(x)$ actually fixed points of $f_a(x)$ (i.e., not prime period 2 points)? If so, are they attracting or repelling? Are any of the fixed points of $f_a^2(x)$ prime period 2 points of $f_a(x)$? If so, are they attracting or repelling?

3. In exploration 28, you showed that $f_3(2/3) = 2/3$ and $f_3'(2/3) = -1$. What is $f_3''(2/3)$? Look at the figure first to determine $f_3''(2/3)$ from the properties of the graph. Now calculate it exactly. How is this different from the second derivative condition in the tangent bifurcation theorem? Why do you think this is important in explaining the bifurcation that you see when looking at the graph of $y = f_a^2(x)$?

Exploration 30. $\boxed{\texttt{IBLdynamics.com}}$ The tool on the website for this exploration looks more closely at the function from exploration 29. Explore how the fixed point moves and the graph itself looks like it is rotating about the fixed point. Focus on the rotation. Do you see how the graph of $y = f_a^2(x)$ "rotates" through the line $y = x$ as a passes through the bifurcation value of $a = 3$?

1. Explain why this property is important for the creation of three fixed points of $f_a^2(x)$ as a passes through the bifurcation value.

2. What can you say about

$$\frac{\partial}{\partial x}\left(f_a^2(x)\right)$$

    for $x < 3$? For $x = 3$? For $x > 3$?

3. What does this exploration imply about the value of

$$\frac{\partial}{\partial a}\left(\frac{\partial}{\partial x}\left(f_a^2(x)\right)\right) = \frac{\partial^2}{\partial a\partial x}\left(f_a^2(x)\right)$$

    evaluated at $a = 3$, $x = 2/3$?

Conjecture

Conjecture 31. Use your answers from above to write the **Period Doubling Bifurcation Theorem.**

Consider the family of dynamical systems

$$x_{n+1} = F_c(x_n)$$

where $F_c : \mathbb{R} \to \mathbb{R}$ is a C^2 function in terms of both x and c. Additionally, assume that there exists a c_0 and x_0 such that

- _____ (fixed point condition),

- _____ (derivative condition), and

- _____ (derivative condition from 30).

Then there exists intervals $I \ni x_0$ and $J \ni c_0$ and a C^1 function $p : I \to J$ with $p(x_0) = c_0$ such that

- _____ (x is a period 2 point and $p(x)$ is a corresponding parameter value), but

- _____ (x is not a fixed point)

for all $x \in I$. Moreover, $p'(x_0)$ ____ and $p''(x_0)$ ____.

Apply

Application 32. When does a period doubling bifurcation occur in the logistic family? Verify the hypotheses of Theorem 5.2.

Application 33. When does a period doubling bifurcation occur in the quadratic family? Verify the hypotheses of Theorem 5.2.

Prove

Theorem 5.2 *Consider the family of dynamical systems*

$$x_{n+1} = F_c(x_n)$$

where $F_c : \mathbb{R} \to \mathbb{R}$ is a C^2 function in terms of both x and c. Additionally, assume that there exists a c_0 and x_0 such that

- $F_{c_0}(x_0) = x_0$,

- $F'_{c_0}(x_0) = -1$, *and*

- $\dfrac{\partial^2}{\partial c \partial x} \left(F_c^2(x) \right) |_{(c_0, x_0)} \neq 0.$

Then there exists intervals $I \ni x_0$ and $J \ni c_0$ and a C^1 function $p : I \to J$ with $p(x_0) = c_0$ such that

- $F_{p(x)}^2(x) = x$, *but*

- $F_{p(x)}(x) \neq x$

for all $x \in I$. Moreover, $p'(x_0) = 0$ and $p''(x_0) \neq 0$.

Instead of proving this theorem, let's spend some time thinking about what it says and some of the meanings of the hypotheses and the conclusions.

Proof 34. In part 3 of exploration 29 above, you noted from the graphs that $(F_{c_0}^2)''(x_0) = 0$. Use the hypotheses of the Period Doubling Bifurcation Theorem to prove that this is true in general.

Proof 35. What do you think the purpose of the third bullet point in the hypothesis of Theorem 5.2 is? In particular, what do you think it says about how the graph changes with c?

The function $p(x)$ in the conclusion does come from the implicit function theorem, but not applied directly. That is basically why we are not proving this theorem here. The last two questions ask you to think about what the conclusions of this theorem mean.

Proof 36. What do the two bullet points in the conclusion tell us about the function $p(x)$? What does $p(x)$ graph in a bifurcation diagram?

Proof 37. What do the derivative conditions on p imply about this graph? Can you sketch the bifurcation diagram from this?

6

Examples of Global Dynamics

6.1 Local Dynamics vs. Global Dynamics

Your work up to this point has been learning the tools to understand what is called *local dynamics*. For example, given a dynamical system, you can compute the fixed points and describe what happens near those fixed points. We don't have tools to define exactly what we might mean by "near" because the implicit function theorem doesn't give us a way to do that.

You've also spent some time learning about bifurcations in families of functions. And while this work is about families of dynamical systems as opposed to a single dynamical system, the work is still local. A tangent bifurcation, for example, tells us about the emergence of fixed points in pairs, but gives us no information about these dynamical systems away from these points. Nor does it tell us about all the bifurcations that might occur in the given family of dynamical systems.

To fully describe the *global dynamics* of a dynamical system, we would like to completely describe what happens to *every* orbit of *every* dynamical system in a family of dynamical systems. This would be truly global information because we would have a complete description of all possible dynamic behaviors. Not surprisingly, it is usually not possible to attain this level of detail.

So what seems reasonable to do?

1. Identify the points whose orbits stay bounded (i.e., don't go off to infinity).

2. Determine what periodic orbits do and do not exist. If possible, describe their stability.

3. Develop a method for describing the kinds of non-periodic orbits that occur.

In short, we would like to catalog all of the different dynamical behaviors that occur, even though we may not be able to pinpoint the exact orbits that exhibit these behaviors.

In this section, we will introduce an incredibly helpful tool for describing and cataloging the global dynamics of some dynamical systems that exhibit dynamics that initially seem beyond description. This tool boils down to describing orbits with infinite sequences of zero's and one's. And these sequences are pretty straightforward to understand.

But before we do that, let's look back on one of our earlier examples and paint a global portrait of its dynamics to get a feel for our goal.

Explore

All of the problems in this section are about the logistic family of dynamical systems

$$x_{n+1} = f_a(x_n) = ax_n(1 - x_n).$$

Our goal here is to describe the fate of every possible initial condition in $\mathbb{R}$, but we don't limit our analysis to the interval $[0, 1]$ as we have done for much of the earlier work. We are not concerned with the population dynamics interpretation of this dynamical system. We are simply thinking about it as a mathematical object, and thus are not limited to only positive values of x.

Application 1. $\boxed{\text{IBLdynamics.com}}$ Start with the parameter interval $0 < a < 1$. Create a graph of the logistic function $y = f_a(x)$ for a value of a between 0 and 1. You can do this using the tool on the website, using graphing software of your choice, or by hand. Be sure to include the line $y = x$ in your figure. You probably want a plotting domain of $-1 < x < 2$ or slightly bigger. Use this graph to answer the following questions. Some of them review properties that you have learned previously. What is different here is that this sequence of exercises guides you in creating a picture that summarizes global dynamics.

For applications 2 through 8, use the graph from application 1.

Exploration 2. What are the fixed points and what are their stabilities?

Exploration 3. Describe the orbits of all initial conditions that are less than the smallest of the two fixed points.

Exploration 4. Describe the orbits of all initial conditions that are between the two fixed points.

Exploration 5. On your graph, locate the two points that are eventually fixed points. There is one that lands on each of the two fixed points.

Exploration 6. Describe the orbits of all initial conditions that are between the fixed point $x = 0$ and the eventually fixed point $x = 1$.

Exploration 7. Describe the orbits of all initial conditions that are between the eventually fixed point $x = 1$ and the other eventually fixed point.

Exploration 8. Describe the orbits of all initial conditions that are greater than the largest of the two eventually fixed points.

Application 9. $\boxed{\texttt{IBLdynamics.com}}$ Repeat applications 2 through 8 with the parameter interval $1 < a < 3$. The tool on the website might be helpful, but you may need a more complete graph to help with initial conditions outside the interval shown there. You can zoom out on the graph on the website or use a different graphing program.

Application 10. You've done it! Color a pair of number lines in some way to summarize the work that resulted from the work you did in applications 1 through 9. Your colors should signify the different types of behaviors that you observed.

6.2 The Logistic Map with $a = 4$ (Part 1)

Now, let's turn to studying the logistic map when $a = 4$. Ultimately, we will need to fill in what happens when a is between $a = 3$ and $a = 4$, but that will come in a later chapter.

Explore

When $a = 4$, the logistic equation is given by

$$x_{n+1} = f_4(x_n) = 4x_n(1 - x_n). \tag{6.1}$$

Some of the work that you did in applications 1 through 9 will be relevant here. However, much of this is quite different, very interesting, and somewhat counterintuitive. Most importantly, the dynamics that you will discover here and the techniques you will use to discover them illustrate some fundamental concepts in dynamical systems.

Understanding the properties of the graph of $y = f_4(x)$ is essential because these ultimately determine the dynamics. Figure 6.1 shows the graph of $y = f_4(x)$. As usual, the line $y = x$ is also shown. There is also something new in this figure: the lines $x = 1$ and $y = 1$ are superimposed in gray. We'll see why shortly.

Exploration 11. Prove that the maximum value of $f_4(x)$ is $y = 1$. Prove that the fixed points are $x = 0$ and $x = 3/4$ and that both of them are repelling.

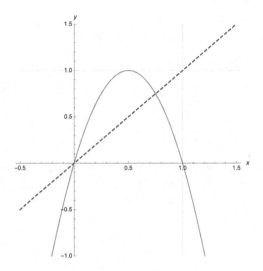

FIGURE 6.1
The graph of the logistic function with $a = 4$.

Exploration 12. Show, using graphical analysis, that the orbits of initial conditions x_0 with $x_0 < 0$ or $x_0 > 1$ tend to negative infinity.

Remember that our ultimate goal is to describe the orbits of **all** initial conditions in $\mathbb{R}$ that occur in equation 6.1. At this point, we are only left with the initial conditions in the unit interval.

Exploration 13. Consider an initial condition $x_0 \in [0, 1]$. Prove that the orbit of this point remains in the interval $[0, 1]$ for all time. More precisely, prove that if $x_0 \in [0, 1]$, then $f_4^n(x_0) \in [0, 1]$ for all $n \geq 0$.

Exploration 14. Use the graph in figure 6.1 to identify points that are eventually fixed to $x = 0$. Identify points that are eventually fixed to $x = 3/4$?

Exploration 15. $\boxed{\texttt{IBLdynamics.com}}$ Print out a copy of this graph (one is available on the website) and perform graphical analysis for several initial conditions $x_0 \in [0, 1]$. Do you observe any periodic behavior? Is there a pattern?

Exploration 16. $\boxed{\texttt{IBLdynamics.com}}$ Repeat exploration 15 using the logistic function iteration tool on the website.

Exploration 17. Thinking about the graph of $y = f_4(x)$, sketch the graph of $y = f_4^2(x)$ on the interval $[0, 1]$ by hand. **Note:** The formula of $f_4^2(x)$ will not help you very much. Instead, think about how the properties of the graph of $y = f_4(x)$ determine the graph of $y = f_4^2(x)$.

Note: The tool on the website for explorations 18 through 23 will plot the graphs of $y = f_4^n(x)$ for $n = 1$ to 7. It will also perform graphical analysis. Use it to check your answer to exploration 17 and then to answer explorations 18 through 23.

Exploration 18. $\boxed{\texttt{IBLdynamics.com}}$ How many critical points does $y = f_4^n(x)$ have? Is there a general formula for the number of critical points in terms of n? What are the critical values (i.e., y-values of these critical points)? Describe the orbits of these points under iteration by f_4.

Exploration 19. $\boxed{\texttt{IBLdynamics.com}}$ What happens to the number of critical points of $y = f_4^n(x)$ and the distance between them as n goes to infinity?

Exploration 20. $\boxed{\texttt{IBLdynamics.com}}$ How many fixed points does $f_4^n(x)$ have? Is there a general formula for the number of fixed points of $f_4^n(x)$ in terms of n? Describe the orbits of these points under iteration by f_4.

Exploration 21. $\boxed{\texttt{IBLdynamics.com}}$ Explain why $f_4(x)$ has periodic points of all periods. Does $f_4(x)$ have prime periodic points of all periods? Why or why not?

Exploration 22. $\boxed{\texttt{IBLdynamics.com}}$ What happens to the distance between fixed points of $y = f_4^n(x)$ as n goes to infinity? What does this tell you about the distance between periodic points of f_4?

Exploration 23. $\boxed{\texttt{IBLdynamics.com}}$ When you set $n = 1$ using the tool on the website and do the graphical analysis, can you find any periodic orbits (other than the obvious fixed points)? Can you find any eventually fixed points (other than the obvious ones)?

Exploration 24. Looking at explorations 18 through 23, can you explain the apparent inconsistencies here? You showed that there are infinitely many eventually fixed points that are arbitrarily close together. You also showed that there are infinitely many periodic points that are arbitrarily close together. This seems to imply that periodic and eventually fixed points are almost everywhere. Yet when you or the computer performs graphical analysis, you don't see any of these orbits! What are some reasons why this might be so?

6.3 The Doubling Map

When you think back on the explorations in section 6.2, you should see that there were two related properties that lead to the infinitude of periodic points (as well eventually fixed points), and to the non-periodic dynamics observed in the stair-step diagrams. The first of these is that the function f_4 maps the interval $[0, 1]$ onto the interval $[0, 1]$. This implies that every orbit remains in the interval $[0, 1]$ for all n. The second property is that the function is "two-to-one." Every point in the interval $[0, 1]$ (except $1/2$) has exactly 2 values that map onto it. We call these two points *preimages*. These two properties combine to cause $f_4^n(x)$ to have $2^n - 1$ critical points with critical values of either 0 or 1, and for there to be 2^n period-n points for all $n > 0$.

Another property of f_4 makes computing these critical points and periodic points difficult, if not impossible. That is its non-linearity. While we can use the quadratic formula to find the fixed points, this tool goes out the window very quickly since $f_4^n(x)$ is a polynomial of degree 2^n.

Clearly we can't construct a continuous, linear, two-to-one function since the graphs of linear functions are lines! However, there is a simple piecewise linear function that has these important properties. This function is called the **doubling map** and is defined by

$$D(x) = 2x \pmod{1} = \begin{cases} 2x & \text{if } 0 \le x < 1/2 \\ 2x - 1 & \text{if } 1/2 \le x \le 1 \end{cases}. \tag{6.2}$$

The graph of the doubling map is shown in figure 6.2. Iterating the doubling map is easy; simply multiply by 2 each time and if the number ever becomes greater than 1, ignore the integer part (that is what the notation mod 1 indicates in equation 6.2). So, for example

$$\frac{5}{12} \mapsto \frac{5}{6} \mapsto \frac{5}{3} \equiv \frac{2}{3} \mapsto \frac{4}{3} \equiv \frac{1}{3} \mapsto \frac{2}{3} \cdots$$

So, $5/12$ is an eventually period 2 point as it lands on the period 2 orbit of $\{1/3, 2/3\}$.

6.3.1   Basic Dynamics of the Doubling Map

Explore

Exploration 25. Describe the orbits of the initial conditions

$$1/4,\ 1/3,\ 1/5,\ 1/6,\ 3/16$$

under iteration by the doubling map D.

Exploration 26. What are the fixed points of D? What are the prime period 2 points? Period 3 points?

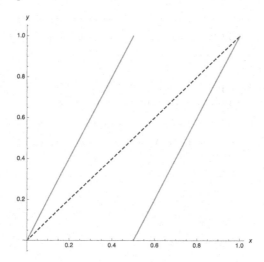

FIGURE 6.2
The graph of the doubling map $D(x)$.

Exploration 27. Describe all eventually fixed points of D.

Exploration 28. Now let's think about the graph of $D(x)$ and its iterates.

1. Carefully plot the graph of $y = D^2(x)$ by hand. How many pieces are there? What are the points of discontinuity? Identify on your graph the period 2 points of D and eventually fixed points of D.

2. Repeat for $y = D^3(x)$.

3. Do you see a pattern here? What can you say about the graph of $y = D^n(x)$. How does this translate to periodic points and eventually fixed points of D?

Exploration 29. $\boxed{\texttt{IBLdynamics.com}}$ The doubling map tool on the website generates stair-step diagrams for D. Do you see any of the periodic behavior that you described above? Why the discrepancy?

Conjecture

Conjecture 30. Complete the following theorem about the number of period n points of the doubling map D.

> The doubling map D has _____ periodic points for each $n > 0$.

Conjecture 31. Complete the following theorem about the distance between consecutive period n points of the doubling map D.

> The distance between consecutive period n points of the doubling map D is less than or equal to_____ for each $n > 0$.

Prove

Theorem 6.1 *The doubling map D has 2^n periodic points for each $n > 0$.*

Proof 32. Prove Theorem 6.1.

Theorem 6.2 *The distance between consecutive period n points of the doubling map D is less than or equal to $1/2^{n-1}$ for each $n > 0$.*

Proof 33. Prove Theorem 6.2.

6.3.2 The Doubling Map in Binary

The best way to fully understand and appreciate the dynamics of the doubling map is to represent the numbers in the unit interval in binary. Some of the material covered here will be review for you, but I encourage you to work through this carefully. The ideas and techniques presented in this section are fundamental to this work and will reappear frequently as we move forward.

Recall that the **geometric series** is

$$1 + x + x^2 + \cdots = \sum_{k=0}^{\infty} x^k = \frac{1}{1-x}, \text{ if } |x| < 1. \tag{6.3}$$

As we will see shortly, the ability to work with geometric series is extremely important in understanding the complicated dynamics that we are beginning to explore. If it has been a while since you worked with geometric series in calculus, I suggest you take some time to go back and review them. One thing to note is that it does matter where the summation index (k in this case) starts and ends. In particular, we will often encounter geometric series that start at $k = 1$ instead of $k = 0$. By starting at $k = 1$, the first term of 1 is omitted from the sum in 6.3 and thus we get

$$\sum_{k=1}^{\infty} x^k = \frac{1}{1-x} - 1 = \frac{x}{1-x}. \tag{6.4}$$

So, why do we need this? We are so comfortable with decimal expansions that we rarely think about what an expression like $0.26723\ldots$ actually means. So, it makes sense to begin reviewing decimal expansions. A decimal expansion $0.d_1d_2\ldots$ is short hand for the series

$$0.d_1d_2\cdots = \sum_{k=1}^{\infty} \frac{d_k}{10^k},$$

where each coefficient d_k takes an integer value between 0 and 9. Take for example the decimal $0.333\ldots$ We know this equals $1/3$ but why? Well, because

$$\sum_{k=1}^{\infty} \frac{3}{10^k} = 3 \sum_{k=1}^{\infty} \frac{1}{10^k} = 3 \left(\frac{1/10}{1-1/10} \right) = \frac{1}{3}. \tag{6.5}$$

Exploration 34. A common misconception is that the decimal $0.999\ldots$ is "really close to 1." Show that in fact $0.999\cdots = 1$.

A binary expansion is almost identical except that the denominators are powers of 2, and the numerators b_k can be only 0 or 1. More formally, a binary expansion is

$$0.b_1b_2\cdots = \sum_{k=1}^{\infty} \frac{b_k}{2^k} \tag{6.6}$$

with $b_k = 0$ or 1. Let's find the value of the binary expansion $s = .010101\ldots$ as an example . We use the same method as we did for the decimal expansion in equation 6.5:

$$s = .010101\cdots = \frac{1}{2^2} + \frac{1}{2^4} + \frac{1}{2^6} + \cdots = \sum_{k=1}^{\infty} \frac{1}{2^{2k}} = \sum_{k=1}^{\infty} \frac{1}{4^k} = \frac{1/4}{3/4} = \frac{1}{3}.$$

It is interesting and helpful to think about what the binary expansion of a number s tells us. It gives us directions to find s in the unit interval. If the first digit is 0, we know that the number is between 0 and $1/2$. If the first digit is 1, then the number is between $1/2$ and 1. I find it helpful to think of 0's as "lefts" and 1's and "rights." The second digit tells us if we are on the left or right side of the interval identified by the first digit. And so on.

Let's return to the example above. Because the first digit of s is 0, we know $s \in [0, 1/2]$. The second digit is 1, so s is in the right half of this interval; that is $s \in [1/4, 1/2]$. The third digit is now 0, so $s \in [1/4, 3/8]$ and so on. This is illustrated in the binary expansion tool on the website.

Exploration 35. $\boxed{\texttt{IBLdynamics.com}}$ This tool shows how we can locate point s through its binary expansion. Each n takes a successive term of the binary expansion to refine the interval further.

Explore

Let's now return to the dynamics of the doubling map, but this time let's do it not with traditional fractions or decimals, but with their binary representation.

Exploration 36. Above, we showed that the binary expansion of $1/3$ is $.010101\ldots$. Write this out as a series and then compute $D(1/3)$ and $D^2(1/3)$. Does this agree with the work you did in exploration 25?

Exploration 37. Repeat exploration 36 except with the initial condition $s = .011010101\ldots$ Review the first example at the beginning of this section where we computed the orbit of $5/12$ under iteration by the doubling map. Are the orbits the same? Use a geometric series calculation to determine the value of s.

Conjecture

Conjecture 38.

Suppose that $s = .b_1 b_2 b_3 \ldots$ in binary. Then $D(s) =$ _____ in binary?

Conjecture 39. What are the fixed points of the doubling map D in binary? The period 2 points? The period 3 points?

$D^n(s) = s$ if and only if _____ .

Conjecture 40. Give several examples (in binary) of eventually fixed points of D. Similarly, give examples of eventually period 2 points of D.

> s is an eventually fixed point of D if and only if _____ .
> s is an eventually period n point of D if and only if _____ .

Conjecture 41. Suppose that the binary representation of a point s is neither repeating nor eventually repeating. What can you say about the orbit of s under D?

Prove

Proof 42. Suppose that a point $s = .b_1b_2b_3 \ldots$ in binary. Prove that $D(s) = .b_2b_3 \ldots$ in binary.

We say that a binary sequence $s = .b_1b_2b_3 \ldots$ is a **periodic binary sequence** of period n if it repeats blocks of length n. For example

$$s = .011011011 \ldots$$

is a period sequence of period 3.

Proof 43. Prove that x is a periodic point of the doubling map D if and only if the binary expansion of x is a periodic binary sequence.

6.4   The Logistic Map with $a > 4$ (Part 1)

The logistic map exhibits some important and interesting behavior when $a > 4$. Remember that our goal is to describe all possible dynamical behavior regardless of initial condition.

Explore

All of the questions in this section are about the iteration of the logistic function when $a = 4.5$ as shown in figure 6.3. As usual, we've included the graph of the line $y = x$ and again have superimposed the lines $x = 1$ and $y = 1$. Printing out at least one copy of figure 6.3 (it is on the website) may be helpful in answering the questions in this section.

Exploration 44. Show, using graphical analysis, that the orbits of points with initial conditions less than 0 or greater than 1 go to negative infinity.

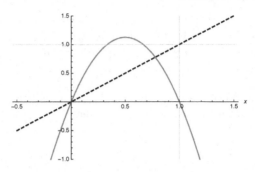

FIGURE 6.3
The graph of the logistic function with $a = 4.5$.

Exploration 45. Now, consider the interval $[0, 1]$. Graphically, identify the values of x in this interval where $f_{4.5}(x) > 1$. What are the orbits of these points under iteration?

Exploration 46. Identify the two points x_L and x_R that are eventually fixed to $x = 0$ in exactly two iterations? How do these points relate to your answer from the previous question?

Exploration 47. Graphically, identify the intervals in $[0, 1]$ that get mapped into the interval (x_L, x_R). What are the orbits of these points under iteration? How many of these intervals are there, and approximately where are they located?

Exploration 48. ┃ `IBLdynamics.com` ┃ Repeat explorations 46 and 47 using the tool on the website for higher iterates. In addition to the graph of the function, the x-axis is color coded to indicate the value of $f_{4.5}^n(x)$. Here are a few things to think about as you do this.

1. How do the intervals relate to properties of the graphs of $y = f_{4.5}^n(x)$ for each n?

2. How many intervals of each color are there for each n? Is there a general formula in terms of n? Why?

3. Are there points in the interval $[0, 1]$ whose orbits remain in that interval for all n? If so, describe them. If not, explain why.

4. Does the process demonstrated here remind you of anything else that you have encountered in mathematics? If so, what?

7

The Tools of Global Dynamics

7.1   How to study Global Dynamics

In chapter 6, you explored dynamical systems that exhibited some surprising properties. In particular, that work suggested that some dynamical systems on $\mathbb{R}$

- have infinitely many periodic orbits,

- have prime periodic orbits of all periods,

- have infinity many eventually fixed points (and presumably eventually periodic points), and

- sometimes exhibit these dynamics on an invariant set having a complicated topological structure.

All of these things and more are true for the examples that you explored in chapter 6. Moreover, these properties are universal. They are the fundamental characteristics of what mathematicians refer to as *chaotic dynamical systems*.

But before we can prove these facts about the examples of chapter 6, we need to take a brief detour to introduce two essential concepts: the *Cantor set* and the *shift map*. We begin with the Cantor middle-thirds set that exemplifies the fractal structure that often accompanies chaotic dynamics. Next, we will introduce a function called the shift map that captures the essence of the dynamics that you discovered in chapter 6.

Once we understand these two tools, we can then use them to prove that the dynamical systems of chapter 6 are in fact chaotic. We begin this process by using the dynamics to understand the structure of the invariant set and in many cases, this allows us to prove that this set has the structure of the Cantor set. We then can use this structure to construct a new function that puts the dynamics of the given dynamical system into correspondence with the dynamics of the shift map. In this way, we can prove that a given dynamical system has the properties listed in the bullet points above. This work will be done in chapter 8.

7.2   The Cantor Set

Hopefully, your answer to part 4 of exploration 48 in chapter 6 is that logistic map with $a > 4$ has a property reminiscent of the Cantor Set. You may have already encountered the Cantor middle-thirds set in an analysis or topology course. However, we should review some of its remarkable properties here not just because they are interesting, but because they provide us a tool for understanding dynamical systems like the ones in chapter 6.

The construction of the classic Cantor middle-thirds set K is an iterative process that begins with the closed interval $[0, 1]$. In the first step of the construction, remove the middle-third open interval $(1/3, 2/3)$. This leaves us with the set

$$K_1 = \left[0, \frac{1}{3}\right] \cup \left[\frac{2}{3}, 1\right].$$

The next step is to remove the open middle-thirds of each of these two intervals. This leaves a set K_2 consisting of 4 closed intervals:

$$K_2 = \left[0, \frac{1}{9}\right] \cup \left[\frac{2}{9}, \frac{3}{9}\right] \cup \left[\frac{6}{9}, \frac{7}{9}\right] \cup \left[\frac{8}{9}, 1\right].$$

We imagine continuing in this fashion infinitely many times. The result of this infinite process is the Cantor middle-thirds set K. Formally, we define

$$K = \bigcap_{n=0}^{\infty} K_n. \tag{7.1}$$

This set has some remarkable properties that you will prove below. Probably, the most confusing result that you will prove is that in one sense the Cantor set is "infinitely small," while in another sense, it is "infinitely large." The exercises below will make this more precise and lead you through the required proofs.

Prove

The first thing we need to do is very basic: we need to show that this infinite process of removing open middle-thirds intervals leaves something in the set K.

Proof 1. Prove that $K \neq \emptyset$.

Now that we know that K is non-empty, we aim to do two different things to get a feel for its size. On the one hand, we want to "count" the number of elements in K. On the other hand, we want to compute the "length" of K.

Proof 2. Let's be sure that we understand the structure of each of the preliminary sets that are used in constructing K.

1. How many intervals are in K_1? What is the length of each of them? What are the endpoints of these intervals?

2. How many intervals are in K_2? What is the length of each of them? What are the endpoints of these intervals?

3. How many intervals are in K_3? What is the length of each of them? What are the endpoints of these intervals?

4. In general, how many intervals are in K_n? What is the length of each of them?

Proof 3. Prove that if x is an endpoint of K_n for some n, then $x \in K$.

A set X is countably infinite (or countable) if the elements in X can be put in one-to-one correspondence with the natural numbers $\mathbb{N}$. One way of doing this is to develop a method or algorithm to list the points in X. Once you have such a list, then you can use it to specify the first point of K, the second point of K, the hundredth point of K, and so on.

Proof 4. Prove that the set of all endpoints of K is countably infinite by developing a procedure to list them.

Proof 5. Prove, using a geometric series, that the total length of the intervals *removed* to construct K sums to 1. What does this imply about the "length" of K? **Hint:** Proof 2 suggests how to set up this series.

Proof 6. Prove that K contains no intervals.

So, now you know that the Cantor set is "really small" in the sense that its total length equals 0. You started out with an interval of length 1 and removed a collection of intervals totaling length 1. So in the sense of length (technically measure), K is "really small."

Proof 7. Prove that $x \in K$ if and only if

$$ x = \sum_{n=1}^{\infty} \frac{s_n}{3^n}, \text{ where } s_n = 0 \text{ or } s_n = 2. \tag{7.2} $$

You might be a little concerned about this last statement since we know that

$$ \frac{1}{3} = \frac{1}{3} + \frac{0}{3^2} + \frac{0}{3^3} + \cdots \in K $$

and this is **not** of the form given in equation 7.2. This is not the problem that it seems at first.

Proof 8. Show that
$$\frac{1}{3} = \sum_{n=2}^{\infty} \frac{2}{3^n}.$$

We can deal with other problematic endpoints of K in a similar fashion.

You probably noticed the similarity between equation 7.2 in proof 7 and the binary expansions discussed in chapter 6. They are both based on geometric series, but in equation 7.2, there is a 3^n in the denominator instead of a 2^n. Expansions such as these, with $s_n = 0$, 1, or 2, are called **ternary expansions.** They can be thought of in the same way we thought of binary expansions except instead of a left and right, we now have a left (0), middle (1), and right (2).

It is reasonable to think that the process of creating the Cantor set only leaves the endpoints of each K_n in K. We also know from proof 4 that the set of endpoints is countable. So, it seems reasonable to guess that the Cantor set K is countably infinite. Surprisingly, this is not the case.

Proof 9. Show that the Cantor set K is uncountable by constructing a function $F : K \rightarrow [0,1]$ that is one-to-one and onto. **Hint:** Use ternary and binary expansions. F is really simple if you do it this way.

Proof 10. Explain why proof 9 implies that there are points in K that are **not** endpoints of intervals removed.

Proof 11. Explain why proof 9 implies that K is "really large" in the sense of counting. Discuss why it is remarkable that K is both "really large" in one sense and "really small" (proof 5) in another. Compare the properties of the Cantor set to other, more common, subsets of $\mathbb{R}$.

7.3 The Shift Map (Part 1)

In section 6.3.2, you learned that the easiest way to describe the global dynamics of the doubling map D is to represent numbers in the interval $[0,1]$ in binary. Doing this allowed you to easily list all of the periodic points and all of the eventually periodic points. It turns out that this idea can be generalized to the other functions that we studied in chapter 6. In this section, we generalize the lessons of section 6.3.2 in preparation for applying them to the other dynamical systems discussed in chapter 6.

7.3.1 The Sequence Space on 2 Symbols

Define the **sequence space on two symbols** by

$$\Sigma_2 = \{.s_0 s_1 s_2 \ldots \mid s_i = 0 \text{ or } 1 \text{ for all } i\}. \tag{7.3}$$

In other words, Σ_2 is a space whose points are all infinite sequences of zeros and ones. We begin by defining a **metric** (i.e., a distance function) d on Σ_2. Let $s = .s_0s_1s_2\ldots$ and $t = .t_0t_1t_2\ldots$ be any two elements of Σ_2. Define a metric $d : \Sigma_2 \to \mathbb{R}$ by

$$d(s,t) = \sum_{k=0}^{\infty} \frac{|s_k - t_k|}{2^k}. \tag{7.4}$$

Metrics, like the one defined here, must satisfy 3 properties:

1. $d(s,t) = d(t,s)$. This is called *commutativity*.

2. $d(s,t) \geq 0$ for all s and t with equality if and only if $s = t$. This is called *positivity*.

3. $d(s,t) \leq d(s,u) + d(u,t)$ for all s, u, and t. This is called *the triangle inequality*.

Explore

Let's begin by computing some distances to get comfortable with Σ_2 and the metric d. Observe that equation 7.4 looks very much like a geometric series, except the numerator is not constant. Because of that, all of these calculations boil down to using the geometric series formula 6.3.

Exploration 12. Let $s = .\overline{0}$ and $t = .\overline{1}$ (the overbar indicates that the sequence repeats). Show that $d(s,t) = 2$. Prove that $d(s,t) \leq 2$ for all s, $t \in \Sigma_2$.

Exploration 13. Let $s = .\overline{01}$ and $t = .\overline{1}$. Compute $d(s,t)$.

Exploration 14. Let $s_1 = .1111\overline{0}$ and $t = .\overline{1}$. Compute $d(s_1,t)$. Let $s_2 = .11111\overline{0}$ and compute $d(s_2,t)$. Which of s_1 and s_2 is nearer to t?

Exploration 15. Let $s_1 = .1010\overline{0}$ and $t_1 = .1010\overline{1}$. Compute $d(s_1,t_1)$. Let $s_2 = .101010\overline{0}$ and $t_2 = .101010\overline{1}$ and compute $d(s_2,t_2)$. Which pair of points are closer to together?

Exploration 16. In general terms, when do you think that two points s and t are "close together"?

Conjecture

Use your observations from explorations 23 through 16 to formalize the notion of "closeness" in Σ_2.

Conjecture 17.

Let $s, t \in \Sigma_2$ and suppose that $s_i = t_i$ for $i = 0, 1, \ldots, n$. Then

$$d(s,t) \leq \underline{\hspace{2cm}}.$$

Conjecture 18.

Let $s, t \in \Sigma_2$. If $d(s,t) < \dfrac{1}{2^n}$ then $\underline{\hspace{2cm}}$.

Apply

Application 19. The ideas presented here can be generalized to more than just infinite sequences of 0s and 1s. Can you define Σ_3, the sequence space on 3 symbols? What is a metric on this space? What about Σ_n?

Application 20. Let $s = .\overline{0}$, $t = .\overline{1} \in \Sigma_N$. Compute $d(s,t)$.

Application 21. Let $0 \leq k \leq n - 2$ and define $s = .\overline{k}$, $t = .\overline{(k+1)} \in \Sigma_N$. Compute $d(s,t)$.

Application 22. Show that if $s, t \in \Sigma_N$ then $d(s,t) \leq N$.

Exploration 23. Let $0 \leq k \leq n - 1$. If $s_1 = .kkkk\overline{0}$ and $t = .\overline{1}$ compute $d(s_1, t)$. If $s_2 = .kkkkk\overline{0}$ compute $d(s_2, t)$. Which of s_1 and s_2 is nearer to t?

Proof 24. Prove that $d(s,t)$, as defined in equation 7.4, is a metric by showing that it has each of the three listed properties.

Proof 25. Prove the following theorem.

Theorem 7.1 *Let s, $t \in \Sigma_2$ and suppose that $s_i = t_i$ for $i = 0, 1, \ldots, n$. Then*

$$d(s,t) \leq \frac{1}{2^n}.$$

Proof 26. Prove the following theorem.

Theorem 7.2 *Let s, $t \in \Sigma_2$. If $d(s,t) < \dfrac{1}{2^n}$, then $s_i = t_i$ for $i = 0, 1, \ldots, n$.*

7.3.2 Dynamics on the Sequence Space on 2 Symbols

Now let's return to dynamical systems, but this time our state space will be Σ_2 instead of $\mathbb{R}$. Define the function $\sigma : \Sigma_2 \to \Sigma_2$ by

$$\sigma(.s_0 s_1 s_2 \ldots) = .s_1 s_2 \ldots.$$

The function σ is called the **shift map** because it simply shifts the sequence left one space and deletes the first entry. For example,

$$\sigma(.011100\ldots) = .11100\ldots.$$

The shift map defines a dynamical system on Σ_2. We iterate it in the same way we have been doing for functions on $\mathbb{R}$. The only difference here is that instead of using the real numbers as the state space, we are using infinite sequences of zeros and ones.

Exploration 27. What are the fixed points of σ?

Exploration 28. What are the period 2 points of σ? Period 3 points? Generalize for period n points?

Exploration 29. Let $s = .1111\bar{0}$ and $t = .\bar{1}$.

1. What is $d(s, t)$ (you answered this in a previous question)? Are they "close" together?

2. What is $d(\sigma(s), \sigma(t))$?

3. What is $d(\sigma^4(s), \sigma^4(t))$? Are they "close" together?

Exploration 30. Let $s = .\bar{0}$. Give an example of a point t such that $d(s, t) < \dfrac{1}{2^5}$ and $d(\sigma^n(s), \sigma^n(t)) = 2$ for $n > 6$.

Exploration 31. Repeat the previous exercise, but this time start with a point s that is a prime period 2 point.

Exploration 32. Give an example of an eventually fixed point. Give an example of an eventually period 2 point.

Exploration 33. Can you think of any other interesting dynamical behavior that this function might have? Give an example and explain what happens and why you think that it is interesting.

Exploration 34. For each n, how many points of period n does σ have? **Note:** Count all period n points, not just those of prime period n.

Exploration 35. Are there prime periodic points of all periods?

Exploration 36. Suppose that I gave you some point s in Σ_2 and some $\varepsilon > 0$. Describe how to find a periodic point p such that $d(s, p) < \varepsilon$?

Prove

While you are familiar with the idea of a continuous function on $\mathbb{R}$, you may not have seen that the idea of continuity extends to any metric space X. The definition of continuity below is almost identical to the definition of continuity that you first learned in calculus. The only difference is that we replace the absolute value (the distance function on $\mathbb{R}$) with the more general distance function d.

Definition 7.1 *Let X be a metric space with metric d. A function $f : X \to X$ is continuous if for every $\varepsilon > 0$ there exists $\delta > 0$ such that if $d(x, y) < \delta$, then $d(f(x), f(y)) < \varepsilon$.*

Proof 37. Prove that σ is continuous.

In mathematics, we often need to describe ways in which one set U relates to another set V. There are a few ways to do this that you already know. Phrases such "subset of" and "intersects" describe basic relationships between different sets. However, sometimes we want to describe a relationship between sets in more detail. Are the points of U only in one part of V, or are they uniformly spread out? Are there lots of points of U in V, or just a few? The concept of density, defined below, is one tool that can be used to answer the first of these questions.

Definition 7.2 *Let X be a metric space with metric d and suppose $U, V \subset X$. A set U is **dense** in a set V if for every $\varepsilon > 0$ and every point $u \in U$, there exists a $v \in V$ such that $d(u, v) < \varepsilon$.*

To illustrate this definition let's begin with proofs involving familiar sets.

Proof 38. Show that the rational numbers are dense in the real numbers.

Proof 39. Show that the set

$$U = \left\{ \frac{1}{n} \ : \ n \in \mathbb{N} \right\}$$

is *not* dense in the interval $[0, 1]$.

Now let's return to our sequence space Σ_2 and the shift map σ.

Proof 40. Prove that periodic points of σ are dense in Σ_2.

Proof 41. Prove that eventually fixed points of σ are dense in Σ_2.

8

Examples of Chaos

8.1 Introduction: The Definition of Chaos

All of the dynamical systems that you have explored in the past two chapters are called *chaotic* dynamical systems. Chaos, in the mathematical sense, was first encountered in the 1960s independently by a variety mathematicians and mathematically inclined scientists. Possibly, the first person to describe this phenomenon was the meteorologist Edward Lorenz in his 1963 paper *Deterministic non-periodic flow* published in the *Journal of Atmospheric Science*. [7] The story of Lorenz's discovery of chaos and the stories of the other pioneers in this field are remarkable. We highly recommend the book *Chaos: The Making of a New Science* by James Gleick [3] for those who may be interested. It vividly tells these stories, describes the underlying mathematical ideas for non-specialists, and explains why chaos was such a surprising and important mathematical and scientific discovery.

Definition 8.1 *A dynamical system $F : X \to X$ is* **chaotic** *if*

- *Periodic points of F are dense in X,*

- *F displays sensitive dependence on initial conditions on X, and*

- *F is topologically transitive on X.*

You should already understand the meaning of the first bullet point in the definition. The other two need definitions and explanations.

Explore

Let's begin with *sensitive dependence*.

Definition 8.2 *A dynamical system $F : X \to X$ displays* **sensitive dependence on initial conditions** *on X if there exists $b > 0$ such that for every $\varepsilon > 0$ and every $x \in X$, there exists a $y \in X$ with $d(x, y) < \varepsilon$ and an $N > 0$ such that $d(F^N(x), F^N(y)) > b$.*

There are a lot of quantifiers to parse in this definition, and we should take some time to get a feel for what it says. In short, it says that F displays sensitive dependence on initial conditions if two close points (x and y), eventually get separated by a distance b under iteration by F. We need to discuss this more carefully to fully understand all of the nuances of the definition. The best way to do this is to slowly and carefully analyze each phrase in the definition and then see how they work together to capture the concept of sensitive dependence.

First, definition 8.2 says "*for every x,*" which means that this phenomenon happens everywhere in X.

Next, it says "*there exists a $b > 0$.*" The constant b is a "separation constant" as it gives us a measure of how far apart the orbits of x and y will become at later time N. This is stated in the conclusion that there exists an N such that $d(F^N(x), F^N(y)) > b$.

Now on to the ε phrase in the definition. This phrase says "*for every ε.*" We should think of ε as a measure of initial closeness. This is expressed in the inequality $d(x, y) < \varepsilon$.

Finally, let's talk about the y. This time the definition says "*there exists a y,*" which implies that the separation might not happen for all points close to x, but only for at least one point close to x.

Hopefully, this careful description of the definition is helpful. However, it is probably best to see what happens when we iterate a system that displays sensitive dependence on initial conditions. The tool on the website for explorations 1 through 5 allows you to explore this phenomenon when iterating the logistic map with $a = 4$.

Exploration 1. ⟦`IBLdynamics.com`⟧ Using the tool on the website, set $\varepsilon = 0.0001$. Try to find an initial condition where the distance between iterates remains less than $1/2$ for all iterates displayed on the tool. Repeat this with $\varepsilon = 0.00001$.

Exploration 2. ⟦`IBLdynamics.com`⟧ Try to find a value of $\varepsilon > 0$, where the distance between orbits is always less than $1/2$ for the 20 iterates shown?

Exploration 3. ⟦`IBLdynamics.com`⟧ If the distance between orbits ever becomes greater than $1/2$, does this distance remain so for all iterates thereafter?

Exploration 4. ⟦`IBLdynamics.com`⟧ Let $x_0 = 0.2$ and $\varepsilon = 0.0001$. Track the orbits by making a list using the letters L and R. Write an L (for left) every time an iterate is less than $1/2$ and write an R (right) every time an iterate is greater than $1/2$.

1. How do these lists relate to one another, if at all?

2. What is the first entry where one list has an L but the other list has an R? Call this value N for future reference.

Exploration 5. $\boxed{\text{IBLdynamics.com}}$ Convert your sequences of letters from Exploration 4 to a sequence of numbers by replacing each L with 0 and each R with 1. Call these sequences s and t, respectively.

1. Find an upper bound for $d(s, t)$ in Σ_2 using the metric given in equation 7.4. This is an estimate on how **close** the points are together initially.

2. Use the value of N that you found in exploration 4 to find a **lower bound** on $d(\sigma^N(s), \sigma^N(t))$. This is an estimate on how far apart these iterates are after N iterates.

3. How do you think this relates to sensitive dependence from definition 8.2?

The third characteristic of chaos, as defined in definition 8.1, is topological transitivity.

Definition 8.3 *A function* $F : X \to X$ *is* **topologically transitive** *if for any pair of open sets* U *and* V *in* X *there exists* $k > 0$ *such that*

$$F^k(U) \cap V \neq \emptyset.$$

This definition basically says that if a dynamical system is topologically transitive on a set X, then the open sets of X get all mixed up under iteration. More specifically, it says that given any two open sets U and V, an iterate of U will eventually intersect V.

To understand definition 8.3, we need to first understand what is meant by $F(I)$ where I is an interval. Formally, we define

$$F(I) = \{y \mid y = F(x), \text{ for all } x \in I\}.$$

The first topological transitivity tool on the web site uses the logistic function $f_4(x)$ to illustrate this basic idea. You define an interval I and the tool displays its image $f_4(I)$.

Exploration 6. $\boxed{\text{IBLdynamics.com}}$ Use the tool on the website to choose the left and right endpoints (xmin and xmax) of an input interval I so that both endpoints have values less than $1/2$.

1. How are the endpoints of the image interval $f_4(I)$ determined?

2. For a point in the image interval, how many points in the domain interval I get mapped onto that point?

Answer these questions again when xmin and xmax are both greater than $1/2$.

Exploration 7. $\boxed{\text{IBLdynamics.com}}$ Now, take xmin $< 1/2 <$ xmax and again use the tool on the website to observe $f_4(I)$.

1. How are endpoints of the image interval $f_4(I)$ determined?

2. For a point in the image interval, how many points in the domain interval I get mapped onto that point?

Exploration 8. The endpoints of the image interval $f_4(x)$ were determined differently in exploration 7 than they were in exploration 6. What caused this difference?

Exploration 9. Your answers to item 2 in explorations 6 and 7 also differed. What caused this difference?

Exploration 10. How do your answers to explorations 8 and 9 relate to each other?

The tool on the website for exploration 11 shows sequences of interval iterates and how these image intervals relate to a fixed target interval.

Exploration 11. [IBLdynamics.com] Play with this tool a bit and try to find examples where the iterated interval does not eventually intersect the target interval. How does this relate to the concept of topological transitivity?

Prove

Directly proving topological transitivity for a given dynamical system can sometimes be tricky. However, for some functions, there is a much easier way to prove this that uses the concept of denseness given in definition 7.2. Recall that for a dynamical system F, the orbit $\mathcal{O}(x)$ is the set of iterates of the initial condition x and is formally defined by

$$\mathcal{O}(x) = \{F^n(x)\}_{n=0}^{\infty}.$$

Theorem 8.1 *Let X be a metric space with metric d. If $F : X \to X$ is continuous and has a dense orbit, then F is topologically transitive.*

The following steps guide you through a proof of this theorem. Let U and V be any two open sets in X. Our goal is to show that there exists $N > 0$ such that

$$F^N(U) \cap V \neq \emptyset.$$

Proof 12. Explain why we can assume that that $U \cap V = \emptyset$.

The assumption of Theorem 8.1 tells us that there exists a point $x \in X$ such that $\mathcal{O}(x)$ is dense in X.

Proof 13. Show that there exists an $n_1 > 0$ such that

$$F^{n_1}(x) \in U.$$

Let $y = F^{n_1}(x)$.

Proof 14. Show that there exists an $n_2 > n_1$ such that

$$F^{n_2}(x) \in V.$$

Proof 15. Show that
$$F^{n_2 - n_1}(y) \in V.$$

Proof 16. Explain why this implies that

$$F^{n_2 - n_1}(U) \cap V \neq \emptyset.$$

8.2 The Shift Map (Part 2)

We now return to the study of the shift map σ and prove that it is chaotic on Σ_2.

Prove

Proof 17. Prove that periodic points of σ are dense in Σ_2.

Proof 18. Prove that σ exhibits sensitive dependence on initial conditions by completing the following steps.

1. Fix $\varepsilon > 0$. Explain why we can choose $N > 0$ such that $\dfrac{1}{2^N} < \varepsilon$.

2. Fix any point $s = .s_0 s_1 \cdots \in \Sigma_2$. Define t so that $d(s,t) < (1/2)^N$. **Hint:** What must the sequence t look like if s and t are close?

3. Further, define t to ensure that

$$d\left(\sigma^N(s), \sigma^N(t)\right) \geq 1.$$

4. Explain why this proves the result.

Proof 19. Consider a point s in Σ_2 constructed in the following manner.

1. Let the first two entries of s be 0 and 1.

2. Append to that all sequences of 0's and 1's having length 2. (there are 4 of them.)

3. Append to that all sequences of 0's and 1's having length 3. (how many are there?)

4. Keep appending sequences in this manner. **Note:** You can't write out s completely, but you should be able to explain the process of doing so in detail.

Prove that the orbit of s under iteration by σ is dense in Σ_2.

Proof 20. Devise a different method for constructing another point s' that also has a dense orbit.

8.3   Topological Conjugacy

Now that you have shown that the shift map σ is chaotic on Σ_2, we want to use that fact to prove that many of the other systems that we have explored are also chaotic. The basic methodology for doing this is to

- construct a function that creates a correspondence between a carefully chosen subset X of $\mathbb{R}$ and Σ_2, and

- use this function to show that the dynamics of the original function are equivalent to the dynamics of the shift map σ. This equivalence is known as a **conjugacy.**

Let S be the function that defines the correspondence between X and Σ_2. Generally, we need S to create a one-to-one correspondence between points of our dynamic set X and points of our sequence space Σ_2 and thus S must be

- one-to-one,

- onto, and

- continuous with a continuous inverse S^{-1}.

A function that has these properties is called a **homeomorphism.** Below are the definitions of one-to-one and onto. You have likely encountered these concepts in other mathematics courses.

Definition 8.4 *A function* $f : X \to Y$ *is* **one-to-one** *if for all* x_1 *and* x_2 *in* X, $f(x_1) = f(x_2)$ *implies that* $x_1 = x_2$. *Equivalently, if* $x_1 \neq x_2$ *then* $f(x_1) \neq f(x_2)$.

Definition 8.5 *A function* $f : X \to Y$ *is* **onto** *if for every* $y \in Y$ *there exists an* $x \in X$ *such that* $f(x) = y$.

A Note on Function Nomenclature

The terminology used by mathematicians to describe different classes of functions might seem a bit confusing to you. This is partly because different mathematical disciplines want to describe different functional properties and this has led to a plethora of similar sounding terms describing different functional properties.

One common nomenclature is to use the suffix "morphism" when describing a function. A one-to-one function is called a **monomorphism**. An onto function is called an **epimorphism**. A function that is both one-to-one and onto is called a **homeomorphism**. A **diffeomorphism** is a homeomorphism that is also differentiable. And the names keep coming.

One important warning is in order. You may have encountered **homomorphisms** in an abstract algebra class. This is a type of function that is unrelated to the *homeomorphisms* that we are using here. It is unfortunate that these two largely unrelated classes of functions are described by words differing in only a single letter.

The method used to establish that two different functions possess equivalent dynamics is summarized in the diagram of figure 8.1 where arrows are used to represent the action of a function. Along the top row, we represent the dynamical function F that we are studying. Along the bottom row is a representation of the shift map σ. The homeomorphism S is represented by the two vertical arrows on the left and right of the diagram. These connect the underlying phase spaces. In this case, S maps values of X to values of Σ_2. Because S is a homeomorphism, we say that the spaces X and Σ_2 are *topologically equivalent*.

Using the homeomorphism S to establish the topological equivalence of the spaces is only half of the problem. We also need to establish *dynamical equivalence* of the functions being iterated. This is done by looking at the interaction between the dynamical functions and the homeomorphism S as represented in the commutative diagram. The diagram is about function composition and the easiest way to understand such a diagram is to follow the functions from the upper left of the diagram to the bottom right of the diagram.

- If we go "right then down," the diagram tells us to first apply F (top arrow right) and then apply S (right arrow down). Thus, that pathway is $S(F(x))$.

- If we go "down then right," the diagram tells us to first apply S (left arrow down) and then apply σ (bottom arrow right). Thus, that pathway is $\sigma(S(x))$.

When this diagram applies, it says that

$$S(F(x)) = \sigma(S(x)). \tag{8.1}$$

$$X \xrightarrow{F} X$$
$$\downarrow S \qquad \downarrow S$$
$$\Sigma_2 \xrightarrow{\sigma} \Sigma_2$$

FIGURE 8.1
A commutative diagram for the topological conjugacy between a function F and the shift map σ.

This relationship, combined with the fact that S is a homeomorphism, establishes the dynamical equivalence between F and σ. This is how one proves that a function F is chaotic.

The hardest part of using this method to establish that a dynamical system is chaotic is not defining the function S, but showing that it satisfies all of the required properties discussed above. Because of that, we will not complete all of those proofs. One good resource for the proofs required is *An Introduction to Chaotic Dynamical Systems* by Robert L. Devaney [2].

Apply

Diagrams such as the one shown in figure 8.1 are known as **commutative diagrams** and are used widely in mathematics where the composition of multiple functions in multiple ways are being compared. We have labeled the upper and lower topological spaces as X and Σ_2 simply because that is what they will be in our context. These spaces could be $\mathbb{R}$, $\mathbb{R}^n$, $\mathbb{C}$ or almost anything else.

Application 21. You have encountered the concept embodied in the commutative diagram above in linear algebra when you studied the concept of *similar matrices*. A matrix A is similar to a matrix B if there exists an invertible matrix P such that

$$B = PAP^{-1}. \tag{8.2}$$

1. Rewrite equation 8.2 so that it is in the form of equation 8.1.

2. What are the spaces at the corners of the commutative diagram in this setting?

3. Draw the commutative diagram expressed in equation 8.2.

Application 22. What linear algebra concept does the "commutative diagram" below represent?

$$\begin{array}{ccc} \vec{v} & \xrightarrow{\ A\ } & \lambda\vec{v} \\ {\scriptstyle P}\downarrow & & \downarrow{\scriptstyle P} \\ \vec{e} & \xrightarrow{\ D\ } & \lambda\vec{e} \end{array}$$

8.4   Return to The Doubling Map

In this section, you will prove the doubling map $D(x) = 2x \pmod 1$ is chaotic on the interval $[0,1]$. Despite the warning at the end of the previous section, we can complete all the tasks needed to do this for this example.

Prove

Proof 23.   Review section 6.3 on the doubling map D. What dynamical properties of D suggest that this function is chaotic on $[0,1]$?

Proof 24.   There is a very natural way to define a function $S : [0,1] \to \Sigma_2$ that is hinted at in the doubling map exercises of section 6.3.2. What is it?

Proof 25.   Prove that S is one-to-one.

Proof 26.   Prove that S is onto.

Proof 27.   Prove that S is continuous. **Note:** This is almost as easy as it seems, but you need to be careful with ε and δ. Also, remember that the domain of S is $[0,1]$ and the range of S is Σ_2.

Proof 28.   Prove that S^{-1} is continuous.

At this point, you have proved that S is in fact a homeomorphism. The final task is to use this homeomorphism to show that the dynamics of D and σ are equivalent.

Proof 29.   Prove that $S(D(x)) = \sigma(S(x))$ and conclude that the doubling map D is chaotic on $[0,1]$.

8.5   The Logistic Map with $a > 4$ (Part 2)

Now its time to return to the logistic map to prove that the logistic map with $a > 4$ is chaotic. It will be helpful to review the material covered in section 6.4 where you first explored these dynamics. Figure 8.2 shows the graphs of $y = f_{4.5}(x)$, (left) $y = f_{4.5}^2(x)$, (center), and $y = f_{4.5}^3(x)$, (right) to help you remember some of the essential properties. In particular, because the central portion of the graph of $y = f_{4.5}(x)$ is above the line $y = 1$, we know that the orbits of points in the center iterate to negative infinity. Similarly, there is an interval to the left of the center and one to the right of the center that get mapped into the center and hence also iterate to negative infinity. In fact, you showed that for each n there are 2^{n-1} open intervals such that if x is in one of these intervals, then

$$f_{4.5}^{2^{n-1}}(x) > 1.$$

Hence, these values of x iterate to negative infinity. This leaves very few points whose orbits remain in the interval $[0, 1]$ for all time.

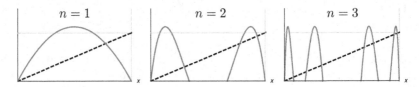

FIGURE 8.2
The graphs of $y = f_{4.5}^n(x)$ with $n = 1, 2, 3$.

Several steps are required to prove that this function is chaotic and it will be helpful to outline the process before working through the actual proofs.

1. You have already shown that almost every point has an orbit that goes to negative infinity. However, there is a set of points (that we will call Λ) whose iterates remain in the unit interval for all time. Our first step will be to determine the topological structure of this set Λ.

2. The second step is to develop a method for describing the orbits of points in Λ. This method (actually a function) is called an *itinerary* because it describes the orbit of a point x by telling us where it is at every time in its orbit.

3. Finally, we will use the itinerary to create a topological conjugacy between the logistic function restricted to Λ and the shift map σ on Σ_2. This will prove that the logistic function is chaotic on Λ.

Prove

Let's begin by introducing some notation. Let

$$I_0 = \left\{ x \in \left[0, \frac{1}{2}\right] \;\middle|\; f(x) \le 1 \right\} \tag{8.3}$$

and

$$I_1 = \left\{ x \in \left[\frac{1}{2}, 1\right] \;\middle|\; f(x) \le 1 \right\}. \tag{8.4}$$

The intervals I_0 and I_1 are highlighted along the x-axis in figure 8.3. Note that $I_0 \cup I_1$ is simply the set of all x-values such that $f(x) \in [0, 1]$. In mathematical notation

$$I_0 \cup I_1 = \{ x \in [0, 1] \mid f(x) \in [0, 1] \}.$$

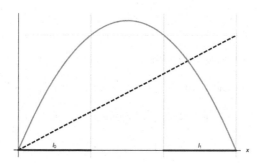

FIGURE 8.3
The graph of $y = f_{4.5}(x)$ with the intervals I_0 and I_1 highlighted.

All of the results that you are about to prove are true when the parameter $a > 4$. However, the proofs are easier if $a > 2 + \sqrt{5}$. We will assume this property for the remainder of this section. The following exercise motivates this choice of a.

Proof 30. Prove that if $x \in I_0 \cup I_1$, then $|f'_a(x)| > 1$. **Hint:** For what values of x is $|f'_a(x)|$ *smallest* for $x \in I_0 \cup I_1$? Look at figure 8.3.

Proof 31. Prove that I_0 has two disjoint subintervals I_{00} and I_{01} such that

- if $x \in I_{00}$ then $f_a(x) \in I_0$ and $f_a^2(x) \in I_0$, and

- if $x \in I_{01}$ then $f_a(x) \in I_0$ and $f_a^2(x) \in I_1$.

Identify these intervals on figure 8.3.

Proof 32. Repeat proof 31 for subintervals of I_1. Specifically, prove that I_1 has two disjoint subintervals I_{10} and I_{11} such that

- if $x \in I_{10}$ then $f_a(x) \in I_1$ and $f_a^2(x) \in I_0$, and

- if $x \in I_{11}$ then $f_a(x) \in I_1$ and $f_a^2(x) \in I_1$.

Identify these intervals on figure 8.3.

Proof 33. We can naturally continue to define subintervals in this way. Describe the orbits of points in I_{001} and I_{101}. How many subintervals of type $I_{s_0 s_1 s_2}$ are there if $s_i = 0$ or 1 for $i = 0, 1, 2$? Where is each located?

Proof 34. In general, how many subintervals of type $I_{s_0 s_1 s_2 \ldots s_k}$ are there if $s_i = 0$ or 1 for $i = 0, 1, \ldots, k$?

Proof 35. What kind of set is being constructed using this process?

Now let's define the invariant set Λ that we are interested in. Let

$$\Lambda = \{x \in [0,1] \mid f_a^n(x) \in [0,1] \ \forall \ n\}. \tag{8.5}$$

Proof 36. Show that Λ is nonempty.

Proof 37. What is the relationship between Λ and $I_0 \cup I_1$?

Proof 38. What is the relationship between Λ and

$$\bigcup_{s_i \in \{0,1\}} I_{s_0 s_1}?$$

Proof 39. In general, what is the relationship between Λ and

$$\bigcup_{s_i \in \{0,1\}} I_{s_0 s_1 \ldots s_n}?$$

Proof 40. Show that Λ contains no intervals by following the steps below.

1. Assume, by way of contradiction, that there exists an interval $[\alpha, \beta] \in \Lambda$. Explain why $[\alpha, \beta]$ is contained in either I_0 or I_1.

2. One implication of the mean value theorem is that if $m = \min |f_a'(x)|$ on $[\alpha, \beta]$, then
$$|f_a(\alpha) - f_a(\beta)| \geq m \, |\alpha - \beta|.$$
   Use this fact to reach a contradiction.

Proof 41. How can we use the subintervals defined in proof 33 to track the orbit of a point in $x \in \Lambda$ using a sequence of 0's and 1's?

Proof 42. Define a function $S : \Lambda \to \Sigma_2$ based on the result of proof 41. $S(x)$ is called the **itinerary** of x.

Proof 43. Assume for the moment that S is a homeomorphism (i.e., one-to-one, onto, and continuous) to answer the following questions. (You will prove some of these properties shortly).

1. Prove that Λ is uncountable and conclude that Λ must be topologically equivalent to the Cantor set.

2. Prove that
    $$S \left(f_a \left(x \right) \right) = \sigma \left(S \left(x \right) \right)$$
    for all $x \in \Lambda$. Conclude that f_a is chaotic on Λ.

As mentioned above, the itinerary function S is a homeomorphism. Verifying some of the required properties is tricky. In the exercises below, we focus on two properties that are fairly straightforward to prove.

Proof 44. Prove that S is one-to-one. **Hint:** Assume that it is not. What must be true about points in the interval connecting two points with identical itineraries? Use the mean value theorem again.

Proof 45. Prove that S is onto. **Hint:** Use the itinerary to create a nested sequence of closed intervals.

8.6   The Logistic Map with $a = 4$ (Part 2)

We complete our exploration into chaotic dynamics by returning to the logistic map $f_4(x)$. You will show that this dynamical system is chaotic on the interval $[0, 1]$ by showing that it is topologically conjugate (technically semi-conjugate) to the doubling map $D(x)$.

Prove

Proof 46. Let $h_1(x) = \cos(2\pi x)$ and $q(x) = 2x^2 - 1$. Show that

$$h_1(D(x)) = q(h_1(x)).$$

Hint: What is the double angle formula?

Proof 47. Draw the commutative diagram like that shown in figure 8.1 to illustrate the result of proof 46.

Proof 48. Find a function $h_2(x) = Ax + B$ such that

$$h_2(q(x)) = f_4(h_2(x)).$$

Proof 49. Append this conjugacy to the commutative diagram that you made in proof 47.

Proof 50. Look at your two-level commutative diagram. What is the conjugacy function between the doubling map D and the logistic function f_4?

Proof 51. Briefly explain why this implies that f_4 is chaotic on $[0, 1]$.

Proof 52. Recall that a conjugacy function must be one-to-one, onto, and continuous. The one that you just created violates one of these properties and that is why it is called a semi-conjugacy. What property is violated and why?

9

From Fixed Points to Chaos

9.1 Introduction

Let's take a few minutes to review what we know about the logistic family of functions. In chapter 3, you showed that the logistic family of functions

- has a fixed point at $x = 0$ and it is attracting if $0 < a < 1$, and

- has a fixed point at $x = (a-1)/a$ and it is attracting if $1 < a < 3$.

You summarized this by creating the plot of the value of the attracting fixed point as a function of the parameter a like the one shown in figure 9.1.

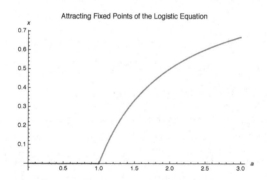

FIGURE 9.1
A fixed point diagram for the logistic family for $0 < a < 3$.

Additionally, in chapter 5 where you studied bifurcations, you learned that at $a = 1$ a tangent bifurcation occurs and at $a = 3$, a period doubling bifurcation occurs. So, this figure neatly summarizes everything we know about the asymptotically stable dynamics of the logistic family for $0 < a < 3$.

Then, in chapter 6, you explored chaotic behavior in great detail and learned that the dynamics of the logistic family of functions is chaotic when $a \geq 4$. So, the natural question to ask is "how do the dynamics go from the very tame dynamics seen when $a < 3$ to incredibly complicated dynamics when $a \geq 4$?" That is the goal of this chapter.

That transition is summarized in a **bifurcation diagram** for the logistic family of functions. Bifurcation diagrams, like the one shown in figure 9.2, are often computed numerically, not analytically. The horizontal axis is the parameter value a and the vertical axis gives a representation of the asymptotic or long term dynamics for each parameter value. For $0 < a < 3$, this figure is a numerically calculated version of figure 9.1. Take a few minutes to convince yourself that this is indeed the case.

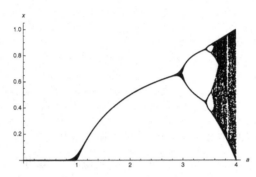

FIGURE 9.2
A bifurcation diagram for the logistic family for $0 < a < 4$.

In this section, you will

- learn how to compute bifurcation diagrams like this,

- learn how to read and interpret the diagram, and

- explore some of the patterns that take a system from equilibrium dynamics to chaotic dynamics.

Remember, the mathematical ideas that we are exploring for the logistic family are ubiquitous! Many other dynamical families have almost identical dynamics and bifurcations and in that sense this diagram is universal.

Unlike in previous chapters, your exercises of the in this chapter will be primarily numerical explorations using apps on the website and conjectures based on these explorations. There are proofs of the ideas that you will explore in this section, but they require some sophisticated mathematical methods. However, there are some remarkable patterns hidden in figure 9.2 to be discovered with a good computer program and a little patience.

9.2 Computing a Bifurcation Diagram

The basic idea behind computing a bifurcation diagram is fairly simple.

1. Make a list of parameter values $\{a_1, a_2, \ldots, a_M\}$.
2. For each parameter value a_k, choose an initial condition x_0 and then plot the $N + 1$ points

$$\{(a_k, x_0), (a_k, x_1), \ldots, (a_k, x_N)\}.$$

This basic outline will need some refinement to get an accurate picture, but it captures the essence of the algorithm. The next few exercises and theorems will tell you how to make these refinements to display an accurate bifurcation diagram.

Explore

Exploration 1. Remember that you only want to see the attracting orbits (when one exists). Modify the plotting of the points in step 2 of the method outlined above so that you are only plotting approximations of the attracting orbits and not the iterates that are "far away" from it?

Exploration 2. Take another look at the bifurcation diagram in figure 9.2. Notice that near the bifurcation values $a = 1$ and $a = 3$, the figure isn't very crisp. There is a thick black smear of points and not a single, fine curve. Why do you think that happened? What could be done to improve this?

A more complicated question is how to pick an initial condition or initial conditions to generate a figure that is an accurate representation of the bifurcation diagram. You want to be sure that the initial condition you choose will converge to an attracting orbit when one exists and that you have accounted for all possible attracting orbits that might exist. In other words, if there are two attracting periodic orbits, we want to see both of them in the bifurcation diagram. It turns out that the critical points of a function are the key to doing this in general. We won't go into detail explaining why this is so (if you are interested in the details, see one of the two Devaney books [1, 2]). However, we will present the important definitions and theorems that tell us this is in fact so.

Definition 9.1 *The Schwarzian derivative of a C^3 function $F : \mathbb{R} \to \mathbb{R}$ is*

$$SF(x) = \frac{F'''(x)}{F'(x)} - \left(\frac{F''(x)}{F'(x)}\right)^2. \tag{9.1}$$

And you always wondered the third derivative might be good for! An important fact related to the iteration of functions is that the Schwarzian derivative remains negative when we compose a function with itself.

Lemma 9.1 *If $SF < 0$, then $SF^n < 0$ for all $n \geq 0$.*

The other important concept that we need to introduce is that of a basin of attraction.

Definition 9.2 *The **basin of attraction** of a fixed point x_0 of a function F is the set of all points whose orbit converges to x_0. The **immediate basin of attraction** of x_0 is the maximal interval containing x_0 that lies in the basin of attraction.*

These two definitions and lemma 9.1 are used to prove the following theorem.

Theorem 9.1 *Suppose $SF < 0$. If x_0 is an attracting periodic point of F, then either the immediate basin of attraction of x_0 extends to $+\infty$ or $-\infty$ or else there is a critical point of F whose orbit is attracted to x_0.*

Prove

Proof 3. Explain why theorem 9.1 implies that a polynomial of degree n with negative Schwarzian derivative can have at most $n + 1$ attracting orbits.

Proof 4. Show that if F is a quadratic polynomial, then $SF < 0$.

Proof 5. Prove the following for the logistic family of functions. If f_a has an attracting periodic orbit Ω in $[0, 1]$, then $f^n(1/2) \to \Omega$ as $n \to \infty$. **Hints:** The points of a period k orbit are fixed points of what function? How many critical points must this function have?

Proof 6. Modify the algorithm for computing the bifurcation diagram of the logistic function using this new idea.

9.3   Period-doubling to Chaos

Now that we understand how to construct bifurcation diagrams like the one shown in figure 9.2, we can begin to explore the details of the diagram and interpret them in terms of dynamics and bifurcations.

Explore

Let's begin our exploration of this bifurcation diagram by looking at what happens to the dynamics as we increase the parameter a past 3 to about 3.5. Figure 9.3 is a magnification (or zoom) of the bifurcation diagram in this region.

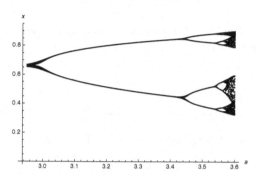

FIGURE 9.3
The bifurcation diagram for the logistic family for $2.95 < a < 3.6$.

Exploration 7. What attracting periodic orbit exists for $3.1 < a < 3.4$?

Exploration 8. What attracting periodic orbit exists for $a = 3.5$?

Exploration 9. Using figure 9.3, describe what you think happens to the periods of the attracting orbits as a increases from 3 to about 3.6.

Exploration 10. | IBLdynamics.com | Use the tool on the website to zoom in on the bifurcation diagram for the logistic function. Start with the parameter range $3.5 < a < 3.6$. What is the highest periodic orbit you can find? You should keep zooming in on this region to explore what is happening in this region.

Exploration 11. | IBLdynamics.com | You know that an attracting period 2 orbit exists for a slightly bigger than 3 because we have shown that a period-doubling bifurcation occurs at $a = 3$. Use the logistic bifurcation tool on the website to answer the following questions.

1. At approximately what value of a does the period 2 orbit bifurcate to a period 4 orbit?

2. At approximately what value of a does the period 4 orbit bifurcate to a period 8 orbit?

3. At approximately what value of a does the period 8 orbit bifurcate to a period 16 orbit?

What is happening to the lengths of these intervals?

You are seeing what is called the period-doubling cascade.

period 1 $\longrightarrow$ period 2 $\longrightarrow$ period 4 $\longrightarrow$ period 8 $\longrightarrow \cdots$.

Exploration 12. Based on these explorations, do you think that this cascade stops at some finite period 2^n or not?

9.4   Windows of Stable Periodic Behavior

Now let's focus on the right half of figure 9.2 past the first period-doubling cascade.

Explore

Exploration 13. IBLdynamics.com Using the tool on the website, give an example of a parameter value where there does not appear to be any attracting periodic orbit. Do you think the dynamics are or are not chaotic?

Exploration 14. IBLdynamics.com Use the tool on the website to find a value of a where there is an attracting period 3 orbit. An attracting period 5 orbit. An attracting period 7 orbit.

You have certainly noticed that in this diagram, there are "windows" of attracting dynamics. Let's look at these a bit more closely.

Exploration 15. IBLdynamics.com Use the tool to zoom in on the period 3 window (use $3.8 < a < 3.87$ at first but you may need to narrow this range as you work through this exercise).

1. What happens to the period 3 orbit as a increases through this range? In other words, what attracting periodic orbits are "descended" from the original periodic orbit.

2. How does this relate to what you observed in section 9.3?

3. Find some other windows of stable periodic behavior and again zoom in on it. What do you notice?

4. Do you have any ideas about why the period 3 orbit "suddenly" appears from the chaotic behavior?

Exploration 16. $\boxed{\texttt{IBLdynamics.com}}$ Let's more carefully explore the emergence of the period 3 window using the graphical analysis tool on the website. You will want to manipulate the parameter a from about $a = 3.8$ to $a = 3.87$. Pay particular attention to what happens to the graph of $y = F^3(x)$ as the period 3 orbit appears. Describe what caused the sudden emergence of this window. In particular, what bifurcation does this look like?

Apply

There is an interesting pattern in the bifurcation diagram that is related to the ordering of the periodic windows in the bifurcation diagram. On the left side is the period 1 window and the biggest window elsewhere in the diagram is the period 3 window. Use the tool on the website to complete the following exercises to discover this pattern.

Application 17. $\boxed{\texttt{IBLdynamics.com}}$ Zoom in on the diagram to see that the biggest two windows between the period 1 and period 3 windows are of periods 5 and 6 and that the order of them is

1 6 5 3.

Application 18. $\boxed{\texttt{IBLdynamics.com}}$ Repeat this zooming in process to fill in this listing.

1 A B 6 C D 5 E 3.

For example, the letter **E** here represents the period of the biggest window between the period 5 and period 3 window.

Application 19. $\boxed{\texttt{IBLdynamics.com}}$ One more time, but now only between periods 1 and 6.

1 a b A c d B e 6.

(Note the **A** and **B** you determined in application 18.)

Application 20. Do you see a pattern? Could you complete one more step in this process without actually zooming in?

10

Sarkovskii's Theorem

10.1   Introduction

One of the overarching goals of this book is to describe, as completely as possible, the entire range of dynamical behavior that can occur in a given dynamical system. One aspect of this goal is to describe the types of periodic behavior that can and cannot occur in a given system. In chapter 6, we did this primarily by using the shift map to show that those dynamical systems that are conjugate to the shift map on Σ_2 have periodic orbits of all periods and that those orbits are dense in the invariant, chaotic set defined by the conjugacy.

The goal of this section is a bit different. We aim to determine what prime periodic orbits must exist if we already know the prime period of one orbit already. You have already done a little of this work in chapter 3. In that chapter, you proved that if f is continuous and there exists a period 2 orbit, then f must also have a fixed point. But what can we say if somehow we know that f has a period 13 point? Is there a period 12 point? What about a period 14 point?

The answer to this question turns out to be quite remarkable. And what makes the answer especially satisfying is that its proof uses no high-powered mathematical machinery. In fact, it relies almost solely on one often overlooked theorem of calculus: the intermediate value theorem.

10.2   The Intermediate Value Theorem

Every result in this section is a consequence of the intermediate value theorem. Informally, the intermediate value theorem says that we can draw a continuous function without lifting our pencil from the paper. This intuition is important, but to understand and apply this theorem, we need a deeper and more formal understanding than the pencil analogy gives us.

Theorem 10.1 (intermediate value theorem) *If $f(x)$ is a continuous function on the interval $[a, b]$ and c is between $f(a)$ and $f(b)$, then there exists an $x \in [a, b]$ such that $f(x) = c$.*

In practice, we often use the following corollaries. The first is simply the case where we assume that $c = 0$ and is actually equivalent to the Intermediate Theorem. In the second corollary, there is an additional hypothesis that guarantees uniqueness.

Corollary 10.1 *Let $f(x)$ be a continuous function on the interval $[a, b]$. If $f(a) \leq 0$ and $f(b) \geq 0$ or vice versa, then there exists an $x \in [a, b]$ such that $f(x) = 0$.*

Corollary 10.2 *Let $f(x)$ be a continuous, strictly decreasing function on the interval $[a, b]$. If $f(a) \geq 0$ and $f(b) \leq 0$, then there exists a unique $x \in [a, b]$ such that $f(x) = 0$.*

Prove

You will prove the intermediate value theorem by completing each of the proofs in this section. More precisely, you will prove Corollary 10.1, which is equivalent to theorem 10.1.

Assume that $f(a) \leq 0 \leq f(b)$. Let $x_0 = a$ and $y_0 = b$. Define

$$m_0 = \frac{x_0 + y_0}{2}.$$

If $f(m_0) = 0$, then $x = m_0$ and we are done. If $f(m_0) < 0$, let $x_1 = m_0$ and $y_1 = y_0$. If $f(m_0) > 0$, let $y_1 = m_0$ and $x_1 = x_0$. Continue in this fashion creating two sequences $\{x_n\}$ and $\{y_n\}$.

Proof 1. Explain why if either $x_n = 0$ or $y_n = 0$, then we are done. Now, assume that this does not happen. Prove that $f(x_n) < 0$ and $f(y_n) > 0$ for all n.

Proof 2. Prove that the sequence $\{x_n\}$ is monotone non-decreasing and that the sequence $\{y_n\}$ is a monotone non-increasing.

Proof 3. Show that sequences $\{x_n\}$ and $\{y_n\}$ are bounded. What does that imply about each of them?

Proof 4. Show that

$$\lim_{n \to \infty} |x_n - y_n| = 0.$$

What does this imply about the limits of the two sequences?

Proof 5. Use the fact that f is continuous to prove Theorem 10.1

Proof 6. Prove corollary 10.2 using this result.

10.3   Review of Two Fixed Point Theorems

In section 4.1, you proved two important fixed point theorems and it may be worthwhile to return to that section and review them. Theorem 10.2 below is a restatement of theorem 4.1 so that it is easily available to you. Theorem 10.3 is new, but its proof follows directly from the results of section 10.2.

Prove

Theorem 10.2 *If f is continuous and $f([a,b]) \supset [a,b]$, then there exists a point $c \in [a,b]$ such that $f(c) = c$.*

Proof 7.   Draw a picture of a function f having the properties described in the hypothesis of the above theorem.

Proof 8.   Prove this theorem using corollary 10.1 of the intermediate value theorem.

Proof 9.   Prove that if f is continuous and f has a prime period 2 orbit, then f has a fixed point.

The following theorem will be quite useful in upcoming work.

Theorem 10.3 *If f is continuous, decreasing, and $f([a,b]) \supset [a,b]$, then there exists a unique point $c \in [a,b]$ such that $f(c) = c$.*

Proof 10.   Draw a picture of a function f having the properties described in the hypothesis of theorem 10.3.

Proof 11.   Prove this theorem using corollary 10.2 of the intermediate value theorem.

10.4   Sarkovskii's Theorem

As you will see shortly, Sarkovskii's theorem [8] answers the question posed at the beginning of this chapter. It provides a way of deducing the existence of different periodic orbits based on knowing the existence of a single periodic orbit.

10.4.1   Discovering Sarkovskii's Theorem

We begin with a sequence of numerical explorations that hint at the power of this remarkable theorem.

Explore

The following exercises look at periodic orbits of the logistic family of functions and serve to motivate Sarkovskii's theorem. Use the tool on the website to explore each of the following questions. Keep a chart that summarizes what prime periodic orbits do and do not exist for each given periodic orbit.

Exploration 12. IBLdynamics.com When $a = 2.5$, the logistic function has a fixed point. Determine whether it has a prime period n point for $n = 1, ..., 8$.

Exploration 13. IBLdynamics.com When $a = 3.175$, the logistic function has a prime period 2 point. Determine whether it has a prime period n point for $n = 1, ..., 8$.

Exploration 14. IBLdynamics.com When $a = 3.834$, the logistic function has a prime period 3 point. Determine whether it has a prime period n point for $n = 1, ..., 8$.

Exploration 15. IBLdynamics.com When $a = 3.5$, the logistic function has a prime period 4 point. Determine whether it has a prime period n point for $n = 1, ..., 8$.

Exploration 16. IBLdynamics.com When $a = 3.739$, the logistic function has a prime period 5 point. Determine whether it has a prime period n point for $n = 1, ..., 8$.

Exploration 17. IBLdynamics.com When $a = 3.628$, the logistic function has a prime period 6 point. Determine whether it has a prime period n point for $n = 1, ..., 8$.

Exploration 18. IBLdynamics.com When $a = 3.70186$, the logistic function has a prime period 7 point. Determine whether it has a prime period n point for $n = 1, ..., 8$.

Exploration 19. IBLdynamics.com When $a = 3.557$, the logistic function has a prime period 8 point. Determine whether it has a prime period n point for $n = 1, ..., 8$.

Exploration 20. Consider the dynamical system

$$x_{n+1} = x_n + \frac{1}{3} \pmod{1}.$$

Show that every point is of period 3 but there does not exist any other prime periodic orbits. Why is this not a contradiction to Theorem 10.2?

Conjecture

The exploration exercises above point toward a more general truth about the existence of certain kinds of periodic points and how they may or may not depend on other types of periodic points. Think about those exercises as you answer the following questions.

Conjecture 21. If you know that f is continuous and has a fixed point, what other prime periodic points, if any, must it also have?

Conjecture 22. If you know that f is continuous and has a prime period 2 point, what other prime periodic points, if any, must it also have?

Conjecture 23. If you know that f is continuous and has a prime period 3 point, what other prime periodic points, if any, must it also have?

Conjecture 24. If you know that f is continuous and has a prime period 4 point, what other prime periodic points, if any, must it also have?

Conjecture 25. If you know that f is continuous and has a prime period 5 point, what other prime periodic points, if any, must it also have?

Conjecture 26. If you know that f is continuous and has a prime period 6 point, what other prime periodic points, if any, must it also have?

Conjecture 27. If you know that f is continuous and has a prime period 7 point, what other prime periodic points, if any, must it also have?

Conjecture 28. If you know that f is continuous and has a prime period 8 point, what other prime periodic points, if any, must it also have?

10.4.2 Using Sarkovskii's Theorem

Before we state Sarkovskii's theorem, we need to introduce the Sarkovskii ordering of the natural numbers which is a bit unusual. When reading statements involving this ordering, I suggest reading the symbol $\triangleright$ as "Sarkovskii greater than." So for example, $3 \triangleright 5$ would be read as "3 is Sarkovskii greater than 5" and this means that 3 comes before 5 in the Sarkovskii ordering. The Sarkovskii ordering is

$$
\begin{array}{cccccccc}
3 & \triangleright & 5 & \triangleright & 7 & \triangleright & \cdots & \\
3 \cdot 2 & \triangleright & 5 \cdot 2 & \triangleright & 7 \cdot 2 & \triangleright & \cdots & \\
3 \cdot 2^2 & \triangleright & 5 \cdot 2^2 & \triangleright & 7 \cdot 2^2 & \triangleright & \cdots & \\
\vdots & & & & & & & \\
\ldots 2^3 & \triangleright & 2^2 & \triangleright & 2 & \triangleright & 1. &
\end{array}
\tag{10.1}
$$

Let's look at this peculiar ordering more carefully. The ordering starts with all of the odd numbers greater than or equal to 3 listed in increasing order (the first line of equation 10.1). Next, come the odd numbers greater than or equal to 3 multiplied by 2 (the second line of equation 10.1) also listed in increasing order. Then, the odd numbers greater than or equal to 3 multiplied by 4 (the third line of equation 10.1). This continues through all positive integer powers of 2. The ordering concludes with the powers of 2 but now in *decreasing* order and thus ending in 1.

Theorem 10.4 (Sarkovskii's theorem) *Suppose that f is a continuous function on $\mathbb{R}$ and has a prime period n orbit. If $n \triangleright m$, then f has a prime period m orbit.*

Sarkovskii, Yorke, and Chaos

Sarkovskii published theorem 10.4 in the *Ukrainian Journal of Mathematics* in 1962 (an English language reprint can be found here [8]). A combination of political and language barriers caused this seminal paper to be unknown to western mathematicians until the late 1970s.

In 1975, the mathematician James Yorke and his graduate student T.Y. Li published a paper entitled "Period 3 implies Chaos" in the *American Mathematical Monthly* [5]. In this paper, they prove, amongst other results, that the existence of a period 3 orbit implies periodic orbits of all periods. In other words, they proved a small part of Sarkovskii's theorem. This paper is also the origin of the term "chaos" to describe the dynamics that you have been studying here. In the book *Chaos* [3], Gleick relates a story of the initial meeting of Sarkovskii and Yorke, and the political and language barriers that unfortunately severely limited interactions between Soviet mathematicians and their colleagues in the United States and Western Europe during that time.

Explore

Exploration 29. Prove that every natural number n appears once and only once in equation 10.1.

Exploration 30. Check your answers to the conjectures 21 through 28. Are they consistent with this theorem?

Apply

Application 31. At the beginning of this chapter, I posed the question "if f is continuous on $\mathbb{R}$ and has a prime period 13 orbit, must it also have a prime period 12 orbit or a prime period 14 orbit?" What is the answer to this question?

Application 32. If f is continuous on $\mathbb{R}$ and has a prime period 13 orbit, describe all periodic orbits it *must* also have. List all periodic orbits it might *not* have.

Application 33. Can a continuous function on $\mathbb{R}$ have a prime period 56 orbit but not a prime period 24 orbit?

Application 34. Can a continuous function on $\mathbb{R}$ have a prime period 24 orbit but not a prime period 56 orbit?

Prove

Before working on a partial proof of Sarkovskii's theorem, you will explore a few consequences of this remarkable theorem and some related ideas.

Proof 35. Use Sarkovskii's theorem to prove that if a continuous function on $\mathbb{R}$ has only finitely many periodic orbits, then all of them are of period 2^k for some k.

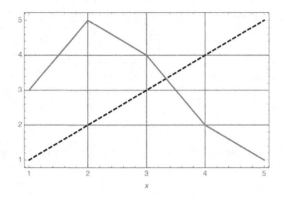

FIGURE 10.1
(Left) The graphs of a function $g(x)$ that has a period 5 orbit but not a period 3 orbit.

In some sense, Sarkovskii's theorem is optimal in that there do exist functions that have a period 5 orbit but not a period 3, or have a period 7 orbit but not a period 5 orbit, and so on. In proofs 36 through 40, you will show that the function illustrated in figure 10.1 has a period 5 orbit, but not a period 3 orbit.

Proof 36. Use figure 10.1 (go to the website for a printable version) to show that the integer values $1, 2, \ldots, 5$ lie on a period 5 orbit of the function $g(x)$. **Note:** Carefully look at the axes! The bottom-left corner is $(1,1)$ not $(0,0)$.

Proof 37. Show that there is not a period 3 point in the interval $[1, 2]$ by finding $g^3([1, 2])$. This should be done using graphical analysis on the interval $[1, 2]$.

Proof 38. Repeat exploration 37 for the intervals $[2, 3]$ and $[4, 5]$.

Proof 39. For the function $g(x)$ in figure 10.1, prove that $g^3([3, 4]) = [1, 5]$. Explain why this implies that there is a fixed point of g^3 in $[3, 4]$.

Proof 40. Use the analysis of the interval $[3, 4]$ done in exploration 39 to show that $g^3([3, 4])$ is *decreasing*. Use that to conclude that the fixed point of g^3 in $[3, 4]$ is not of prime period 3. What is it?

As you might imagine from the strangeness of the Sarkovskii ordering, the complete proof of this statement requires considering many different scenarios. You will tackle a more manageable piece of the proof. You will prove that the existence of a prime period 3 orbit implies the existence of prime periodic orbits of all periods.

Suppose that f is a continuous function with a prime period 3 orbit $\{a, b, c\}$ with $a < b < c$ and $f(a) = b$, $f(b) = c$, and $f(c) = a$.

Proof 41. Draw a number line and place a, b, and c on the line. Draw arrows to indicate how f maps these points to each other. Label the closed interval $[a, b]$ as I_0 and the closed interval $[b, c]$ as I_1.

1. What is $f(I_0)$?
2. What is $f(I_1)$?

We can now begin proving that there exists periodic orbits of all periods. All of this work will use the theorems you proved in the section 10.3 and thus ultimately the intermediate value theorem 10.1.

Proof 42. Prove that there exists a fixed point of f in I_1.

Proof 43. Prove there exists a period 2 point of f in I_0. Be sure to explain why this point cannot be a fixed point of f.

Now, let's turn to proving the existence of a period 4 point.

Proof 44. Let $A_0 = I_1$. Show there exists an interval $A_1 \subset I_1$ such that $f(A_1) = A_0$.

Proof 45. Show there exists an interval $A_2 \subset A_1$ such that $f(A_2) = A_1$. What is $f^2(A_2)$?

Proof 46. Show there exists an interval $A_3 \subset A_2$ such that $f(A_3) = I_0$.

Proof 47. Show that $f^4(A_3) = I_1$. Explain why this implies there is a fixed point of f^4 in A_3?

Proof 48. Why must this fixed point of f^4 be a prime period 4 point of f?

Proof 49. Discuss how you might generalize the argument of proofs 44 through 48 to prove that f has a prime period n orbit for all $n > 3$.

11

Dynamical Systems on the Plane

11.1 Linear Algebra Foundations

Now that we have a good understanding of dynamical systems on the real line, we can use those ideas to study higher dimensional dynamical systems. In this chapter, we focus on dynamical systems in the plane. However, much of what you will learn in this chapter generalizes to higher dimensions. We need to begin by reviewing some important ideas from linear algebra and multivariable calculus that provide a foundation for studying dynamics in $\mathbb{R}^2$.

Let A be a 2×2 matrix with real entries. Then A is a representation of a function called a *linear transformation* from $\mathbb{R}^2$ to $\mathbb{R}^2$. More specifically, if $(x, y) = \mathbf{x} \in \mathbb{R}^2$, then the function $L : \mathbb{R}^2 \to \mathbb{R}^2$ defined by

$$L(\mathbf{x}) = A\mathbf{x} \tag{11.1}$$

is a linear transformation. In section 11.2, you will begin exploring the dynamics of linear transformations on $\mathbb{R}^2$. In this section, we highlight the properties of linear transformations that will aid you in that work.

The most important properties of the matrix A in the study of dynamical systems are its *eigenvalues* and *eigenvectors*.

Definition 11.1 *If A is an $n \times n$ matrix, then λ is an **eigenvalue** of A with associated **eigenvector** $\mathbf{v}$ if*

$$L(\mathbf{v}) = A\mathbf{v} = \lambda\mathbf{v}. \tag{11.2}$$

Note that equation 11.2 is similar to a fixed point equation where the eigenvector $\mathbf{v}$ is "almost fixed" by the linear transformation L. We say that it is almost fixed because the eigenvalue λ is essentially a scaling factor telling us by how much the eigenvector $\mathbf{v}$ gets lengthened or shortened. We will need to make this description more precise because the eigenvalues may or may not be real numbers. But for now, it begins to explain why eigenvectors are important in understanding linear dynamical systems.

Because this is not a linear algebra course, we won't discuss all of the important properties of eigenvalues and eigenvectors and what they tell us about the linear transformation. Below is a short summary of the relevant facts that we will need here.

1. Eigenvalues are found by solving the equation

$$\det(A - \lambda I) = 0$$

   for λ. If A is an $n \times n$ matrix, then this is a degree n polynomial and hence there are n eigenvalues counting multiplicity. This polynomial is called the **characteristic equation** of A.

2. Because the eigenvalues are found by finding the roots of a polynomial, eigenvalues may be real or complex numbers. If the matrix A has real entries, then complex eigenvalues come in complex conjugate pairs $\alpha + i\beta$ and $\alpha - i\beta$.

3. For each eigenvalue λ_i, solve the equation

$$(A - \lambda_i I)\mathbf{v}_i = 0$$

   for a *non-zero* vector $\mathbf{v}_i$. This is an eigenvector associated with the eigenvalue λ_i. When done correctly, there will be at least one free variable to choose. This means that eigenvectors are not unique.

4. If the eigenvalue λ is a root of multiplicity one of the characteristic equation, then any two associated eigenvectors are scalar multiples of each other.

5. Eigenvectors associated with complex eigenvalues also come in complex conjugate pairs. So if $\alpha + i\beta$ has an eigenvector of $\mathbf{u} + i\mathbf{w}$, then $\alpha - i\beta$ has an eigenvector of $\mathbf{u} - i\mathbf{w}$.

6. If the eigenvalues of A are distinct, then the associated set of eigenvectors $B = \{\mathbf{v}_1, \ldots, \mathbf{v}_n\}$ are linear independent and form a basis for the underlying vector space.

7. If the eigenvalues of A are distinct, then the matrix P, whose columns are the eigenvectors, diagonalizes A via the equation

$$P^{-1}AP = D$$

   where D is a diagonal matrix whose diagonal entries are the eigenvalues λ_i. We say that the matrices A and D are *similar*. An important consequence of this is that if A is a diagonal matrix, then its diagonal entries are its eigenvalues.

Eigenvalues, Eigenvectors, and Linear Transformations

The relationship between a matrix and a linear transformation is a little bit more complicated than I let on. In particular, equation 11.1 is imprecise. However, it will suffice for how we are using linear algebra to study dynamical systems. Let me at least hint at the true nature of this relationship here.

A linear transformation is a special kind of function from a vector space V to another vector space W. In our situation, $V = W$. A matrix A is a representation of that linear transformation. There are many different representations of the same linear transformation and these depend on what basis one chooses for the vector space. The eigenvalues are properties of the linear transformation and as such, they are the same for all matrix representations. By contrast, the eigenvectors are properties of the matrix representation. This is why similar matrices have the same eigenvalues, but different eigenvectors. They represent the same linear transformation, but in different bases.

Let's do a simple example. Let

$$A = \begin{pmatrix} 3 & 14 \\ 0 & -4 \end{pmatrix}.$$

The first step is to compute the characteristic polynomial

$$\det(A - \lambda I) = \begin{vmatrix} 3 - \lambda & 14 \\ 0 & -4 - \lambda \end{vmatrix} = (\lambda + 4)(\lambda - 3).$$

This implies that the eigenvalues are $\lambda_1 = -4$ and $\lambda_2 = 3$.

The next step is to compute the eigenvectors for each of these eigenvalues. Let's consider $\lambda_1 = -4$. We need to solve

$$(A - \lambda_1 I)\mathbf{v}_1 = \begin{pmatrix} 7 & 14 \\ 0 & 0 \end{pmatrix} \begin{pmatrix} a \\ b \end{pmatrix} = \begin{pmatrix} 0 \\ 0 \end{pmatrix}$$

for a and b to compute this eigenvector. This is equivalent to the system of equations

$$\begin{aligned} 7a + 14b &= 0 \\ 0 &= 0. \end{aligned}$$

Note that the matrix equation has reduced to essentially one equation and two unknowns. Thus, we have a free choice of either a or b. If we choose $a = 2$, then $b = -1$ and thus our eigenvector is $\mathbf{v}_1 = (2, -1)^T$.

There are two things to note here. First, we are not allowed to choose $a = 0$ because that forces $b = 0$ and and then $\mathbf{v}_1 = \mathbf{0}$. This is not an eigenvector because eigenvectors must be non-zero. Second, you might have made a different choice for a than I did. You might have chosen to let $a = 1$, which forces $b = -1/2$ giving $\mathbf{v}_1 = (1, -1/2)^T$. But if we multiply your eigenvector by 2 we get mine. This is what is meant when we say that eigenvectors are unique up to scalar multiplication.

Exploration 1. Find the eigenvector $\mathbf{v}_2$ in the example above.

Exploration 2. Find the eigenvalues and eigenvectors of the matrix

$$B = \frac{1}{2}\left(\begin{array}{cc} -1 & 7 \\ 7 & -1 \end{array} \right).$$

11.2 Linear Systems with Real Eigenvalues

A linear dynamical system on $\mathbb{R}^2$ has the form

$$
\begin{aligned}
x_{n+1} &= ax_n + by_n \\
y_{n+1} &= cx_n + dy_n
\end{aligned}
$$

which can be expressed as the matrix equation

$$\mathbf{x}_{n+1} = A\mathbf{x}_n \tag{11.3}$$

where $\mathbf{x}_n = (x_n, y_n)^T$ and

$$A = \left(\begin{array}{cc} a & b \\ c & d \end{array} \right).$$

Our goal in this section is to explore the types of dynamics that occur in equation 11.3. Naturally, the dynamics will depend in some way on the entries of the matrix A and to keep things relatively straightforward we will initially consider matrices A that have real eigenvalues. We consider matrices with complex eigenvalues in section 11.3.

Explore

Exploration 3. Show that $\mathbf{0}$ is a fixed point of equation 11.3.

Exploration 4. Show that $\mathbf{x}_n = A^n\mathbf{x}_0$. In other words, show that iteration of the function is simply repeated matrix multiplication.

In the next few exercises, you will look at several examples that illustrate the most common types of dynamics that can occur in equation 11.3. You should use what you have learned about linear systems on $\mathbb{R}$ to describe the dynamics of these planar systems.

Exploration 5. Let

$$A = \begin{pmatrix} .5 & 0 \\ 0 & .2 \end{pmatrix}.$$

1. Rewrite the dynamical system $\mathbf{x}_{n+1} = A\mathbf{x}_n$ as a system of equations in terms of the coordinates x_n and y_n.

2. Describe the orbit of the initial condition $(x_0, 0)$ for all $x_0 \in \mathbb{R}$.

3. Describe the orbit of the initial condition $(0, y_0)$ for all $y_0 \in \mathbb{R}$.

4. Use these results to describe the orbit of the initial condition (x_0, y_0) for all x_0 and $y_0 \in \mathbb{R}$.

5. Draw a picture in the (x, y) plane that illustrates the dynamics of this system. Your picture should reflect all of the dynamical properties that you discovered above.

Exploration 6. Repeat exploration 5 with

$$A = \begin{pmatrix} 1.5 & 0 \\ 0 & 1.2 \end{pmatrix}.$$

Exploration 7. Repeat exploration 5 with

$$A = \begin{pmatrix} 1.5 & 0 \\ 0 & .2 \end{pmatrix}.$$

The pictures that you sketched in the three explorations 5 through 7 are called **phase portraits**. Like the graphical analysis figures that you used to study real dynamical systems, these pictures illustrate the dynamics of the system. However, because we are now working in two dimensions, we are no longer able to use the graph of the function being iterated to help us understand the dynamics. The best we can do is to plot the iterates as points in the (x, y) plane.

These three examples illustrate three fundamental behaviors that occur in linear systems. The phase portrait of exploration 5 illustrates an attracting fixed point sometimes called a **sink**. Exploration 6 is a repelling fixed point or a **source**. Finally, Exploration 7 has one attracting direction (the y-axis) and one repelling direction (the x-axis). A fixed point such as this is called a **saddle**.

Exploration 8. Let

$$A = \begin{pmatrix} -.5 & 0 \\ 0 & .2 \end{pmatrix}.$$

The only difference between this matrix and that given in exploration 5 is that the first non-zero entry is negative. How does this change the dynamics? In particular, is the origin still a sink? What would happen if both entries were negative? What would happen in the other examples if we changed the signs?

In every example above, the axes are invariant in the sense that if an initial condition is on one of these lines, then the orbit of that point remains on that line. This motivates the following definition.

Definition 11.2 *Consider a linear dynamical system* $\mathbf{x}_{n+1} = L(\mathbf{x}_n)$ *with* $L :$ $\mathbb{R}^2 \to \mathbb{R}^2$ *a linear transformation. A line* ℓ *is* **invariant** *if for every* $x_0 \in \ell$, $L^n(x_0) \in \ell$ *for all* $n \geq 0$.

Exploration 9. Let

$$A = \begin{pmatrix} 3/2 & 1 \\ 0 & 2 \end{pmatrix}.$$

1. Describe the orbit of $(x_0, 0)$ for all $x_0 \in \mathbb{R}$. In other words, describe the orbits of points starting on the x-axis.

2. Describe the orbit of $(x_0, 3x_0/2)$ for all $x_0 \in \mathbb{R}$. In other words, describe the orbits of points starting on the line $y = 3x/2$.

3. Is the origin a sink, saddle, or source?

4. In the (x, y) plane, sketch the two lines above and indicate with arrows whether points on them go toward the origin or away from it. Use this picture and your intuition to describe orbits of other initial conditions.

5. Compute the eigenvalues and eigenvectors of the matrix A. How do these relate to the lines and arrows you just graphed?

Exploration 10. Let

$$A = \begin{pmatrix} 1/2 & 1 \\ 0 & 1/3 \end{pmatrix}.$$

1. Describe the orbit of $(x_0, 0)$ for all $x_0 \in \mathbb{R}$. In other words, describe the orbits of points starting on the x-axis.

2. Describe the orbit of $(x_0, -x_0/6)$ for all $x_0 \in \mathbb{R}$. In other words, describe the orbits of points starting on the line $y = -x/6$.

3. Is the origin a sink, saddle, or source?

4. In the (x, y) plane, sketch the two lines above and indicate with arrows whether points on them go toward the origin or away from it. Use this picture and your intuition to describe orbits of other initial conditions.

5. Compute the eigenvalues and eigenvectors of the matrix A. How do these relate to the lines and arrows you just graphed?

Triangular Matrices

You probably noticed that in the previous two exercises, the eigenvalues of A are also the diagonal entries of A. Hopefully, you also realized that this is true because one of the two off-diagonal entries of the matrix equals zero. Matrices where all of the entries below (or above) the diagonal are equal to zero are called **triangular matrices.** If a matrix A is triangular, then the eigenvalues of A are the diagonal entries of A.

If the matrix is not triangular, then the eigenvalues are not the diagonal entries!

Conjecture

The exercises above point to a general theorem about linear dynamical systems with real eigenvalues.

Conjecture 11.

Let A be a 2×2 matrix with real, distinct eigenvalues λ_1 and λ_2 with corresponding eigenvectors $\mathbf{v}_1$ and $\mathbf{v}_2$. Consider the dynamical system

$$\mathbf{x}_{n+1} = A\mathbf{x}_n.$$

• The lines through the origin defined by the eigenvectors $\mathbf{v}_1$ and $\mathbf{v}_2$ are
_____ .

• If _____ then the origin is a sink.

• If _____ then the origin is a source.

• If _____ then the origin is a saddle.

Apply

For each of the matrices A in applications 12 through 16, consider the dynamical system

$$\mathbf{x}_{n+1} = A\mathbf{x}.$$

There are three tasks for each of these applications.

- find the eigenvalues and eigenvectors of A,

- classify the phase portrait as a sink, source, saddle, or neither, and

- sketch the phase portrait including the invariant lines.

Application 12. $A = \begin{pmatrix} 3/8 & 1/8 \\ 1/8 & 3/8 \end{pmatrix}$

Application 13. $A = \begin{pmatrix} 5/6 & 7/6 \\ 7/12 & 17/12 \end{pmatrix}$

Application 14. $A = \begin{pmatrix} 8 & -6 \\ 3 & -1 \end{pmatrix}$

Application 15. $A = \begin{pmatrix} 6 & -6 \\ 1 & -1 \end{pmatrix}$

Application 16. $A = \begin{pmatrix} 5/3 & -4/3 \\ -2/3 & 7/3 \end{pmatrix}$

Prove

Theorem 11.1 *Let A be a 2×2 matrix with real, distinct eigenvalues λ_1 and λ_2 having corresponding eigenvectors $\mathbf{v}_1$ and $\mathbf{v}_2$. Consider the dynamical system*

$$\mathbf{x}_{n+1} = A\mathbf{x}_n. \tag{11.4}$$

- *The lines through the origin defined by the eigenvectors $\mathbf{v}_1$ and $\mathbf{v}_2$ are invariant.*

- *If $|\lambda_i| < 1$ for $i = 1, 2$, then the origin is a sink.*

- *If $|\lambda_i| > 1$ for $i = 1, 2$, then the origin is a source.*

- *If $|\lambda_1| < 1 < |\lambda_2|$, then the origin is a saddle. Moreover, the eigenvector $\mathbf{v}_1$ defines the attracting direction and the eigenvector $\mathbf{v}_2$ defines the repelling direction.*

The following exercises will lead you through a proof of theorem 11.1. Let's begin by formally defining a "line through the origin defined by a vector $\mathbf{v}$." Intuitively, we understand this to mean the line through the origin that extends in the direction of $\mathbf{v}$. Precisely, a line $\ell(\mathbf{v})$ defined by a vector $\mathbf{v}$ is the set

$$\ell(\mathbf{v}) = \left\{ (x, y) \in \mathbb{R}^2 \mid (x, y) = t\mathbf{v} \text{ for some } t \in \mathbb{R} \right\}.$$

Proof 17. Prove that if an initial condition $\mathbf{x}_0$ of equation 11.4 is a member of $\ell(\mathbf{v}_i)$, then there exists a $C \in \mathbb{R}$ such that

$$\mathbf{x}_1 = C\mathbf{v}_i.$$

Conclude that $\mathbf{x}_n \in \ell(\mathbf{v}_i)$ for all $n \geq 0$.

Recall from linear algebra that the **modulus** or **norm** of a vector $\mathbf{v}$ is defined by

$$|\mathbf{v}| = \left| \begin{pmatrix} v_x \\ v_y \end{pmatrix} \right| = \sqrt{v_x^2 + v_y^2}.$$

Proof 18. Suppose that $|\lambda_i| < 1$ and that $\mathbf{x}_0 \in \ell(\mathbf{v}_i)$. Prove that

$$\lim_{n \to \infty} \mathbf{x}_n = \mathbf{0}.$$

This proves that the eigenvectors associated with real eigenvalues less than one in absolute value define attracting invariant lines. Now let's do a similar proof for eigenvalues greater than one in absolute value.

Proof 19. Suppose that $|\lambda_i| > 1$ and that $\mathbf{x}_0 \in \ell(\mathbf{v}_i)$, $\mathbf{x}_0 \neq \mathbf{0}$. Prove that

$$\lim_{n \to \infty} |\mathbf{x}_n| = \infty.$$

Proof 20. Explain why these last three proofs prove theorem 11.1.

There is more that we can say about the orbits of initial conditions on an invariant line. We can, in fact, easily compute the sequence of iterates of these points.

Proof 21. Suppose that λ_i and $\mathbf{v}_i$ are real eigenvalues and eigenvectors of A and that $\mathbf{x}_0 \in \ell(\mathbf{v}_i)$. Prove that

$$\mathbf{x}_n = \lambda_i^n \mathbf{x}_0.$$

Now let's turn our attention to initial condition that are not on either invariant line.

Proof 22. Explain why for all $\mathbf{x}_0 \in \mathbb{R}^2$ there exists unique constants C_1 and C_2 such that

$$\mathbf{x}_0 = C_1 \mathbf{v}_1 + C_2 \mathbf{v}_2.$$

Proof 23. Prove that

$$\mathbf{x}_n = C_1 \lambda_1^n \mathbf{v}_1 + C_2 \lambda_2^n \mathbf{v}_2.$$

Proof 24. Prove if $|\lambda_i| < 1$ for $i = 1, 2$ then for all initial conditions $\mathbf{x}_0$

$$\lim_{n \to \infty} |\mathbf{x}_n| = \mathbf{0}.$$

In other words, prove that if the origin is a sink, then all orbits converge to it.

Proof 25. Suppose that $|\lambda_i| > 1$ and that $\mathbf{x}_0 \neq \mathbf{0}$. Prove that

$$\lim_{n \to \infty} |\mathbf{x}_n| = \infty.$$

This explains why we call the origin a source.

Proof 26. Why can't we make a similar statement when $|\lambda_1| < 1 < |\lambda_2|$? Why do you think this is called a saddle?

11.3 Linear Systems with Complex Eigenvalues

Because eigenvalues of 2×2 matrices are found by finding the roots of a quadratic polynomial, they may either be real or come in complex conjugate pairs. We will see that complex eigenvalues lead to somewhat different dynamics. The goal of this section is to begin the process of understanding the dynamical similarities and differences of linear systems with complex eigenvalues as compared to those with real eigenvalues.

Recall that a number z is complex if $z = a + ib$ where we define $i = \sqrt{-1}$. A quadratic polynomial with real coefficients that has one complex root of $a + ib$ must have a second root of $a - ib$. We say that $a + ib$ and $a - ib$ are **complex conjugates.** If $z = a + ib$, then we denote the complex conjugate of z by $\bar{z} = a - ib$.

Complex or Imaginary?

In my experience, students often confuse the terms *complex* and *imaginary*. A number $z = a + ib$ is a complex number. The value a is called the **real part of** z and the value b is referred to as the **imaginary part of** z. If $z = ib$, then we say that z **is pure imaginary.** In short, $3 + 4i$ is a complex number but $5i$ is pure imaginary.

Explore

Suppose that we want to construct a linear dynamical system

$$\begin{aligned}
x_{n+1} &= ax_n + by_n \\
y_{n+1} &= cx_n + dy_n
\end{aligned}$$

that rotates every point in the plane by some fixed angle ϕ. How do we choose the constants in equation 11.5 to accomplish this goal? Given that we want to rotate every point, it seems natural to represent our initial points in polar form so that

$$\begin{aligned} x_n &= r\cos\theta \\ y_n &= r\sin\theta \end{aligned}$$ (11.5)

for some fixed values of r and θ. We want the point (x_n, y_n) to be rotated by an angle of ϕ so that means that we need

$$\begin{aligned} x_{n+1} &= r\cos(\theta + \phi) \\ y_{n+1} &= r\sin(\theta + \phi). \end{aligned}$$ (11.6)

Exploration 27. Look up the sum of angles formulas for $\sin(\theta + \phi)$ and $\cos(\theta + \phi)$ and rewrite equation 11.6 using them.

Exploration 28. Use equation 11.5 to eliminate both r and θ from your answer to exploration 27 and replace them with x_n and y_n.

Exploration 29. Finally, express this as the matrix equation

$$\begin{pmatrix} x_{n+1} \\ y_{n+1} \end{pmatrix} = \begin{pmatrix} \cos\phi & -\sin\phi \\ \sin\phi & \cos\phi \end{pmatrix} \begin{pmatrix} x_n \\ y_n \end{pmatrix}.$$ (11.7)

Exploration 30. ⏐ IBLdynamics.com ⏐ Use the tool on the website to plot the dynamics of equation 11.7 with

$$\phi = \pi/4, \ \phi = \pi/2, \ \phi = \pi, \ \phi = \pi/10, \ \phi = \pi/\sqrt{2}$$

and answer the following questions for each.

1. Do the dynamics that you see depend on the initial condition?
2. Is the fixed point $(0,0)$ a sink, a source, a saddle, or none of these?
3. Are the dynamics periodic or not? If they are what is the period and how does this relate to ϕ?

Linear systems like the ones you've just explored are called **centers**. Points simply rotate around the origin. These orbits do not tend toward the origin nor away from the origin as n tends to infinity.

Exploration 31. Show that the eigenvalues of the matrix in equation 11.7 are

$$\lambda = \cos\phi \pm i\sin\phi.$$

The **modulus** of a complex number $z = \alpha + i\beta$, denoted $|z|$, is defined by

$$|z| = \sqrt{\alpha^2 + \beta^2}$$

and generalizes the idea of absolute value on $\mathbb{R}$. To see this note that if $z = \alpha + 0i$, then $|z| = |\alpha|$.

Exploration 32. Show that $|\lambda| = 1$ for the eigenvalues of exploration 31.

Let's build on what you just learned and consider the dynamical system

$$\begin{pmatrix} x_{n+1} \\ y_{n+1} \end{pmatrix} = \begin{pmatrix} \rho \cos \phi & -\rho \sin \phi \\ \rho \sin \phi & \rho \cos \phi \end{pmatrix} \begin{pmatrix} x_n \\ y_n \end{pmatrix} \qquad (11.8)$$

where the parameter $\rho > 0$. Note that when $\rho = 1$, equation 11.8 reduces to equation 11.7.

Exploration 33. $\boxed{\texttt{IBLdynamics.com}}$ Use the tool on the website to draw a variety of phase portraits for systems of the form given in equation 11.8.

1. Describe the dynamics when $0 < \rho < 1$.

2. Is the fixed point $(0,0)$ a sink, a source, a saddle, or none of the above?

3. Describe the dynamics when $\rho > 1$.

4. Is the fixed point $(0,0)$ a sink, a source, a saddle, or none of the above?

The orbits that you constructed in exploration 33 are called either **spiral sinks** or **spiral sources** depending on whether the orbit spirals in toward the origin or away from the origin.

Exploration 34. Compute the eigenvalues of the matrix in equation 11.8. What is their modulus?

Conjecture

Conjecture 35.

Consider the linear dynamical system on $\mathbb{R}^2$ defined by

$$\mathbf{x}_{n+1} = A\mathbf{x}_n$$

and let A have complex eigenvalues $\lambda_i = \alpha \pm i\beta$.

- If _____, then $(0,0)$ is a spiral sink.

- If _____, then is a spiral source.

- If _____, then $(0,0)$ is a center.

Apply

For each of the matrices A in applications 36 through 40, consider the dynamical system

$$\mathbf{x}_{n+1} = A\mathbf{x}.$$

There are three tasks for each of these applications.

- find the eigenvalues of A,

- classify the phase portrait as a spiral sink, spiral source, or center, and

- sketch the phase portrait including the invariant lines (feel free to use one of the tools on the website for this).

Application 36. $A = \begin{pmatrix} 3 & -2 \\ 2 & 3 \end{pmatrix}$

Application 37. $A = \begin{pmatrix} 7/3 & 10/3 \\ -4/3 & 11/3 \end{pmatrix}$

Application 38. $A = \begin{pmatrix} 2 & 4 \\ -2 & -2 \end{pmatrix}$

Application 39. $A = \begin{pmatrix} -3/4 & -10/3 \\ 1/3 & 5/4 \end{pmatrix}$

Application 40. $A = \begin{pmatrix} 3 & 10 \\ -1 & -3 \end{pmatrix}$

Prove

The work that you've done in this section points to an interesting fact: complex eigenvalues correspond to rotational behavior in a dynamical system. If λ_1 and λ_2 are complex conjugate pairs of eigenvalues of a dynamical system, then they can be written so that

$$\lambda_i = \rho(\cos\phi \pm i\sin\phi)$$

where $\rho = |\lambda_i|$ is the modulus of λ_i and the angle ϕ is the **argument** of λ_i. The argument ϕ determines the amount of rotation in each iteration. The modulus ρ is the rate of contraction or expansion and if $0 < \rho < 1$, then the iterates tend toward the origin. We thus call this a **spiral sink**. If $\rho > 1$, then the iterates tend away from the origin and we call this a **spiral source**. We call the case where $\rho = 1$ a **center**.

The exercises in the apply section above show that this relationship between complex eigenvalues and the dynamic behavior is true even if the matrix A is not of the form given in equation 11.8. The proof of this fact is not hard but relies on some linear algebra ideas that are not discussed above and are generally not covered in an undergraduate linear algebra course. The basic idea is that a 2×2 real matrix with complex eigenvalues is similar to a matrix of the form given in equation 11.8. This similarity is analogous to the similarity of a matrix A with real distinct eigenvalues to a diagonal matrix D.

All of the results on linear dynamical systems in $\mathbb{R}^2$ discussed in this chapter are summarized in the theorem below.

Theorem 11.2 *Consider the linear dynamical system on $\mathbb{R}^2$ defined by*

$$\mathbf{x}_{n+1} = A\mathbf{x}_n$$

and let A have eigenvalues λ_1 and λ_2.

- *If $|\lambda_i| < 1$ for $i = 1$, 2, then $(0,0)$ is a sink. Moreover, if the imaginary part of λ_i is not equal to 0, then $(0,0)$ is a spiral sink.*

- *If $|\lambda_i| > 1$ for $i = 1$, 2, then $(0,0)$ is a source. Moreover, if the imaginary part of λ_i is not equal to 0, then $(0,0)$ is a spiral source.*

- *If $|\lambda_1| < 1 < |\lambda_2|$, then $(0,0)$ is a saddle.*

- *If $|\lambda_i| = 1$ for $i = 1$, 2 and the imaginary part of λ_i is not equal to 0, then $(0,0)$ is a center.*

11.4   Fixed Points of Nonlinear Systems

Now, let's turn to nonlinear dynamical systems. Let $(x, y) = \mathbf{x} \in \mathbb{R}^2$. Then a function $F : \mathbb{R}^2 \to \mathbb{R}^2$ has the form

$$F(x, y) = F(\mathbf{x}) = (f(x, y), g(x, y)). \tag{11.9}$$

We then define dynamical systems in the usual way by

$$\mathbf{x}_{n+1} = F(\mathbf{x}_n). \tag{11.10}$$

As usual, fixed points are found by solving $F(\mathbf{x}) = \mathbf{x}$. This reduces to solving the system on equations

$$\begin{aligned} f(x, y) &= x \\ g(x, y) &= y \end{aligned} \tag{11.11}$$

for x and y simultaneously. There may be multiple solutions to these equations and hence multiple fixed points for the dynamical system 11.10.

Explore

Let's consider a model of a predator-prey system to illustrate some of the techniques used to analyze a nonlinear dynamical system in the plane. Denote the prey population at time n by x_n and the predator population by y_n. Assume that in the absence of predators, the prey population grows logistically. When predators are present, the prey are harvested at a rate proportional to both the predator and prey population sizes. On the other hand, we assume that in the absence of prey, the predator population decreases at a rate proportional to the predator population size. When prey are present, the predator population increases at a rate proportional to both the predator and prey population sizes. One possible model that satisfies these requirements is

$$\begin{aligned} x_{n+1} &= ax_n(1 - x_n) - \beta x_n y_n \\ y_{n+1} &= \tfrac{4}{5}y_n + 3\beta x_n y_n \end{aligned} \qquad (11.12)$$

where $a > 0$ and $\beta \geq 0$. Because this is a population model, we will assume that $x_0 \geq 0$ and $y_0 \geq 0$.

Let's start by simply connecting this system of equations with the notation used in equation 11.9. In this example, $\mathbf{x}_n = (x_n, y_n)$ so that

$$\begin{aligned} f(x,y) &= ax(1 - x) - \beta xy \\ g(x,y) &= \frac{4}{5}y + 3\beta xy, \end{aligned}$$

and

$$F(x,y) = F(\mathbf{x}) = (f(x,y), g(x,y)).$$

Let's look at what happens when either the predator or prey population is absent.

Exploration 41. Show that if $y_0 = 0$, then $y_n = 0$ for all n. Similarly, show that $x_0 = 0$ implies that $x_n = 0$ for all n. Discuss why this property is reasonable in this model. Explain why this property implies that the x and y axes are invariant?

For the remainder of the exercises in this section, let $a = 2$.

Exploration 42. Use the techniques of one-dimensional dynamical systems to describe the dynamics on each of the axes. In particular, show that $(0,0)$ and $(^1/_2, 0)$ are fixed points.

1. Is $(0,0)$ attracting or repelling for initial conditions on the x-axis?

2. Is $(^1/_2, 0)$ attracting or repelling for initial conditions on the x-axis?

3. Is $(0, 0)$ attracting or repelling for initial conditions on the y-axis?

4. Draw the first quadrant of the (x, y) plane and use arrows to indicate the dynamics on each of these axes.

Next, we turn our attention to what happens when neither x nor y equals zero. For simplicity, we will take $\beta = {}^1/_2$ in what follows.

Exploration 43. Show that $(^2/_{15}, {}^{22}/_{15})$ is the only non-zero fixed point of equation 11.12.

Let's take a break here to review what we know about this dynamical system and what we still need to do yet.

- We know that there are three fixed points at $(0, 0)$, $(^1/_2, 0)$, and $(^2/_{15}, {}^{22}/_{15})$.

- It seems that the fixed point at $(0, 0)$ is a saddle since initial conditions on the y-axis have orbits that converge to it, while initial conditions on the x-axis have orbits that tend away from it.

- We know that the fixed point at $(^1/_2, 0)$ is attracting in one direction (along the x-axis). We don't know what happens to initial conditions that start near $(^1/_2, 0)$ but off the x-axis.

- We know nothing about the fixed point at $(^2/_{15}, {}^{22}/_{15})$.

Our primary goal is to close these gaps. Let's begin this process by numerically computing solutions and noting what we see.

Exploration 44. ⟨IBLdynamics.com⟩ The tool on the website models this predator-prey system. Use it to answer the following questions.

1. There is a point in the tool that can be moved to choose a variety of initial conditions. Do this and describe what you see. In particular, how would you characterize the fixed point at $(^2/_{15}, {}^{22}/_{15})$ using the terminology of theorem 11.2?

2. Use the movable point to choose initial conditions near the fixed point at $(^1/_2, 0)$ and describe these orbits. How would you characterize the fixed point at $(^1/_2, 0)$ using the terminology of theorem 11.2?

3. Use the movable point to choose initial conditions very near the y-axis (especially with $y \leq 1/2$) and describe these orbits. How would you characterize the fixed point at $(0, 0)$ using the terminology of theorem 11.2?

Conjecture

In the predator-prey model of equation 11.12, you found three fixed points. Two of them were saddles and the third was a spiral sink. Based on this, you probably realize that all of the fixed point classifications described in theorem 11.2 for linear systems also occur in nonlinear systems. This should not be a surprise. We saw that this is the case for non-linear dynamical systems on $\mathbb{R}$. We showed that fixed points of linear dynamical systems on $\mathbb{R}$ of the form

$$x_{n+1} = ax$$

are attracting if $|a| < 1$ and are repelling if $|a| > 1$. For nonlinear systems of the form

$$x_{n+1} = f(x_n)$$

we have an almost identical condition for a fixed point $\tilde{x}$. Except in that setting, it is the value $|f'(\tilde{x})|$ that determines whether $\tilde{x}$ is attracting or repelling. In short, *the derivative evaluated at the fixed point determines stability.*

The same thing holds for dynamical systems on $\mathbb{R}^2$. The derivative of F evaluated at the fixed point usually determines the nature of the fixed point. But what is the derivative of a function $F : \mathbb{R}^2 \to \mathbb{R}^2$? You learned in multivariable calculus that the derivative of a function $F : \mathbb{R}^2 \to \mathbb{R}^2$ is a 2×2 matrix called the *Jacobian matrix*. This matrix is denoted DF. The Jacobian matrix consists of four partial derivatives of F and is given by

$$DF = \begin{pmatrix} \frac{\partial f}{\partial x} & \frac{\partial f}{\partial y} \\ \frac{\partial g}{\partial x} & \frac{\partial g}{\partial y} \end{pmatrix}. \tag{11.13}$$

If $\tilde{\mathbf{x}}$ is a fixed point of F, then $DF(\tilde{\mathbf{x}})$ determines the stability of $\tilde{\mathbf{x}}$.

Conjecture 45.

Suppose that $F : \mathbb{R}^2 \to \mathbb{R}^2$ is differentiable and that $F(\tilde{\mathbf{x}}) = \tilde{\mathbf{x}}$.

- If _____, then the $\tilde{\mathbf{x}}$ is a sink.

- If _____, then the $\tilde{\mathbf{x}}$ is a source.

- If _____, then the $\tilde{\mathbf{x}}$ is a saddle.

Apply

Let's return to the predator-prey model we have been exploring.

Application 46. Compute the Jacobian matrix DF for equation 11.12.

1. Find $DF(0,0)$ and compute its eigenvalues to verify that $(0,0)$ is a saddle.

2. Find $DF(1/2,0)$ and compute its eigenvalues to verify that $(1/2,0)$ is a saddle.

3. Find $DF(2/15, 22/15)$ and compute its eigenvalues to verify that $(2/15, 22/15)$ is a spiral sink. (We suggest using a computer algebra system for this calculation.)

Application 47. $\boxed{\texttt{IBLdynamics.com}}$ Now do another investigation of equations 11.12 but this time with $\beta = 1/5$.

1. Begin by using the predator-prey app on the website to simulate this dynamical system to get a feel for what happens.

2. Explain why there are still fixed points at $(0,0)$ and $(1/2,0)$ and why the dynamics on each axis is identical to the previous model.

3. Find the value of the third fixed point. How has reducing the value of β affected the equilibrium predator and prey populations? Interpret these changes in terms of the model.

4. Compute DF and evaluate it at this new fixed point. Find the eigenvalues. How has the stability changed if at all?

Application 48. Find the fixed points and compute their stability.

$$\begin{aligned} x_{n+1} &= 2x_n + 4y_n - 23 \\ y_{n+1} &= -2x_n - 2y_n + 21 \end{aligned}$$

Application 49. Find the fixed points and compute their stability.

$$\begin{aligned} x_{n+1} &= x_n^2 + y_n \\ y_{n+1} &= 2x_n - 3y_n \end{aligned}$$

Prove

Proving that the eigenvalues of the Jacobian matrix $DF(\bar{\mathbf{x}})$ determine the stability of a fixed point is beyond the scope of this text. Below is a statement of the theorem that justifies the use of this matrix to classify fixed points.

Theorem 11.3 *Consider the dynamical system on* $\mathbb{R}^2$ *defined by*

$$\mathbf{x}_{n+1} = F(\mathbf{x}_n)$$

and suppose that $F(\tilde{\mathbf{x}}) = \tilde{\mathbf{x}}$. *Let* $DF(\tilde{\mathbf{x}})$ *have eigenvalues* λ_1 *and* λ_2.

- *If* $|\lambda_i| < 1$ *for* $i = 1, 2$, *then* $\tilde{\mathbf{x}}$ *is a sink. Moreover, if the imaginary part of* λ_i *is not equal to 0, then* $\tilde{\mathbf{x}}$ *is a spiral sink.*

- *If* $|\lambda_i| > 1$ *for* $i = 1, 2$, *then* $\tilde{\mathbf{x}}$ *is a source. Moreover, if the imaginary part of* λ_i *is not equal to 0, then* $\tilde{\mathbf{x}}$ *is a spiral source.*

- *If* $|\lambda_1| < 1 < |\lambda_2|$, *then* $\tilde{\mathbf{x}}$ *is a saddle.*

11.5   Periodic Points

Basically, everything that you already know about periodic points carries over to dynamical systems on $\mathbb{R}^2$. A point $\mathbf{x}_0$ is a period n point if

$$F^n(\mathbf{x}_0) = \mathbf{x}_0.$$

The orbit of $\mathbf{x}_0$ is the set of points $\{\mathbf{x}_0, \mathbf{x}_1, \ldots, \mathbf{x}_{n-1}\}$ with $F(\mathbf{x}_i) = \mathbf{x}_{i+1}$ and $F(\mathbf{x}_{n-1}) = \mathbf{x}_0$.

It is also true that the stability of this orbit is determined by the product of the derivatives evaluated along the periodic orbit. But here we need to be careful. For functions on the real line, these derivative values are scalars and scalar multiplication is commutative. Thus, we don't need to worry about the ordering of the multiplication. However, this is not true on $\mathbb{R}^2$. **Matrix multiplication is not commutative!** If $\{\mathbf{x}_0, \mathbf{x}_1, \ldots, \mathbf{x}_{n-1}\}$ is a periodic orbit, then the stability is determined by the eigenvalues of

$$DF(\mathbf{x}_{n-1}) \cdot DF(\mathbf{x}_{n-2}) \cdots DF(\mathbf{x}_1) \cdot DF(\mathbf{x}_0) \qquad (11.14)$$

in this order.

Apply

As you might imagine, equation 11.14 suggests that it is often quite difficult to compute the stability of a periodic orbit of a planar dynamical system. However, the predator-prey model that we have been working with does give us an opportunity to use this idea in a setting that is fairly straightforward. So, consider equations 11.12 with $a = 3.2$ and $\beta = 0.5$.

Application 50. Use what you already know about the logistic equation to show that there exists a period 2 orbit of the form $\{(x_0, 0), (x_1, 0)\}$ with $0 < x_0 < x_1 < 1$. Explain why this orbit is attracting when restricted to the x-axis. (You don't need to compute the exact values of x_i.)

Application 51. Compute $DF(\mathbf{x}_i, 0)$ and show that it is an upper triangular matrix for each i.

Application 52. Show, in general, that the product of two upper triangular matrices is upper triangular. Observe that the diagonal entries of the product matrix are the products of the diagonal entries.

Application 53. Recall that the eigenvalues of an upper triangular matrix T are the diagonal entries. Let λ_1 be the eigenvalue in the upper-left entry of T and let λ_2 be the eigenvalue in the lower-right entry of T. Show that one of the eigenvectors of T is $(1, 0)^T$. To which eigenvalue does this correspond to? Show that the other eigenvector of T has a non-zero second entry.

Application 54. Now let's put all of this together.

1. Compute $M = DF(\mathbf{x}_1, 0) \cdot DF(\mathbf{x}_0, 0)$.

2. Explain why the eigenvalue of M having the eigenvector $(1, 0)^T$ is less than 1 in absolute value.

3. Show that the other eigenvalue of M is greater than one.

4. Conclude that this period 2 orbit is a saddle.

The following exercises concern an interesting dynamical system that produces something called the "gingerbread attractor" (see figure 11.1). The gingerbread attractor is generated by the dynamical system

$$\begin{aligned} x_{n+1} &= 1 - y_n + |x_n| \\ y_{n+1} &= x_n. \end{aligned} \qquad (11.15)$$

We won't be studying this system in any great detail, but it is a fun example that illustrates some of the basic ideas of this section.

Application 55. Show that the point $(1, 1)$ is a fixed point. Compute the Jacobian matrix $DF(1, 1)$ and find the eigenvalues λ_i. Show that $|\lambda_i| = 1$.

Application 56. Show that the point $(0, 0)$ lies on a period 6 orbit. Identify the points of this orbit in figure 11.1.

Application 57. Show that the point $(-1, -1)$ lies on a period 5 orbit. Identify the points of this orbit in figure 11.1.

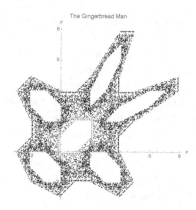

FIGURE 11.1
The Gingerbread attractor.

Application 58. |IBLdynamics.com| Use the tool on the website to explore the orbits of points inside the middle hexagon (the "heart" of the gingerbread man). Describe these orbits.

Application 59. |IBLdynamics.com| Use the tool on the website to explore the orbits of points inside the other hexagons (the "limbs" of the gingerbread man). Describe these orbits. You might try to prove some of this as well. The dynamics inside the "heart" would be a good place to start.

11.6   Chaos in the Hénon map

There is much we have left out of the story on dynamical systems in the plane. For example, there was no discussion of bifurcations. But in this section, we would like to briefly introduce the chaos part of the story as that most closely parallels our study of one-dimensional dynamics. You will study chaos in planar dynamical systems more carefully in chapter 12. There you will build on the symbolic dynamics work of chapter 7 to develop a fuller picture of what you will explore in this section.

The story of chaotic dynamics in the plane begins with a closer look at saddle fixed points. In section 11.4, you learned that when a linear system is a saddle, then there exists a stable invariant line E_s and an unstable invariant line E_u. When a nonlinear dynamical system has a saddle fixed point, it turns out that there exists something quite similar. These are called **stable and unstable curves** and are denoted by W_s and W_u, respectively. Locally, the stable and unstable curves are tangent to the stable and unstable lines E_s and E_u defined by the Jacobian matrix DF evaluated at the fixed point. If you've taken a differential equations course, this is identical to what happens for saddle equilibrium points in that setting.

However, if we look globally, the stable and unstable curves of a fixed point can interact in a surprising way. In some systems, the stable and unstable curves can "cross" each other (this can't happen in differential equations), and this leads to a remarkable phenomenon known as a **homoclinic tangle**. Homoclinic tangles are the hallmark of chaos in planar dynamical systems. And although we will not discuss homoclinic tangles in any great detail, the explorations below point at the complexity of this phenomenon in a dynamical system known as the *Hénon map*.

Explore

The Hénon map is given by

$$
\begin{aligned}
x_{n+1} &= 1 + y - ax^2 \\
y_{n+1} &= bx.
\end{aligned}
\tag{11.16}
$$

We will take $a = 1.4$ and $b = 0.3$.

Exploration 60. ⟨IBLdynamics.com⟩ Use the tool on the website allows you to iterate the Hénon map. Describe in your own words what orbits are converging to. This object is known as the Hénon attractor.

Exploration 61. Show that the Hénon map has a fixed point at approximately

$$(\tilde{x}, \tilde{y}) = (0.631355, 0.189405).$$

Exploration 62. Compute the Jacobian matrix of the Hénon map and evaluate it at $(\tilde{x}, \tilde{y})$. Compute the eigenvalues and show that this fixed point is a saddle. **Note:** This problem should not be done by hand.

Recall that a primary reason that the logistic function is chaotic is that it is a two-to-one function. A consequence of this is that it is not invertible. In two dimensions, there is more "real estate" that allows a function to smoothly stretch and fold a region while still being one-to-one. Each of the next two exercises shows that the Hénon map is invertible and hence one-to-one. Thus, something quite different is happening here that causes the apparent chaotic behavior that you observed in exploration 60.

Exploration 63. Our first method of showing that the Hénon map is invertible uses the inverse function theorem (4.4). Show that if $b \neq 0$, then the determinant of the Jacobian matrix of the Hénon map is non-zero for all x and y. Explain why this implies that the Hénon map is invertible.

Exploration 64. Let $H(x, y)$ be the Hénon map of equation 11.16. Show that when $b \neq 0$,

$$H^{-1}(x, y) = \begin{pmatrix} y/b \\ x - 1 + ay^2/b^2 \end{pmatrix}.$$

The fact that H is invertible and that we can compute H^{-1} gives us a method of approximating the stable curve W_s and the unstable curve W_u. The following outlines a method to approximate W_u.

1. Make a list of initial conditions near the equilibrium point $(\tilde{x}, \tilde{y})$ on E_u. We know these points are close to the unstable curve W_u since W_u is tangent to E_u at $(\tilde{x}, \tilde{y})$.

2. For each of these initial conditions, compute a list of iterates.

3. Plot all of these points.

We approximate W_s in an almost identical fashion. The first change is that instead of taking initial conditions on E_u, we take them from E_s. The second is that instead of iterating H, we iterate H^{-1}.

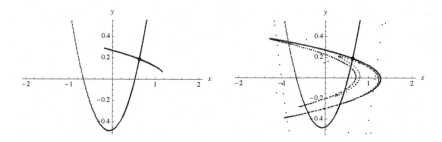

FIGURE 11.2
Images showing a homoclinic tangle in the Hénon map. The plot on the left gives a local picture of W_s and W_u. The plot on the right is a more global picture.

Figure 11.2 shows the result of this method. The plot on the left is the "local" picture. The parabolic shaped curve is approximately the stable curve W_s and the shorter, more horizontal curve is approximately W_u. Note that W_s and W_u intersect at the fixed point. The figure on the right is the "global picture" and gives a glimpse of the homoclinic tangle. Again, the parabolic shaped curve is an approximation of W_s. Note that now W_u looks very much like the Hénon attractor that you observed in exploration 60. To generate this figure, we just computed a few more iterates of each initial condition than we did for the figure on the left. In the "global figure," we see multiple intersections of W_s and W_u. This is the appearance of the homoclinic tangle that creates the attractor and the associated chaotic dynamics.

One final observation: If we take any small box centered at a saddle point, each iteration stretches that box (in the unstable direction E_u) and compresses that box (in the stable direction E_s). The Hénon map adds an additional aspect to this transformation - a folding. If you look at an image of the Hénon attractor you will note that, while not quite symmetric with respect to the x-axis, it does appear to be folded across the x-axis. This leads to one final question for you.

Exploration 65. What property of the Hénon map of equation 11.16 might be responsible for this folding across the x-axis?

12

The Smale Horseshoe

12.1 Motivating the Horseshoe Map

We ended chapter 11 with a discussion of the Hénon map (equation 11.16), homoclinic tangles, and saddles. We tried to imply in that discussion that somehow the Hénon map H stretched, compressed, and folded the plane in a way that created the Hénon attractor and chaos. The goal of this chapter is to look at a topological model of this process and to use symbolic dynamics to prove that this process does, in fact, lead to chaos.

Before doing this in any detail, let's reflect back on two specific examples of one-dimensional chaotic dynamics that we studied earlier. The doubling map $D(x)$ and the logistic function $f_4(x)$ both exhibit chaotic behavior on the interval $[0,1]$. If you think back on your study of those examples, you might see that the primary reason that both of these functions are chaotic is that they are two-to-one functions on the entire unit interval. The fact that almost every point in the unit interval has 2 preimages somehow causes the interval to get so mixed up under iteration that the dynamics on this interval are chaotic.

Let's try and visualize how each of these two functions acts on the unit interval. The action of the doubling map D can be viewed like this.

Step 1 Stretch the unit interval to twice its length.

Step 2 Cut this new interval in half.

Step 3 Place these two intervals back into the unit interval.

Step 4 Repeat, . . .

The action of f_4 can be described similarly.

Step 1 Stretch the unit interval to twice its length.

Step 2 Fold this new interval in half.

Step 3 Place this folded interval back into the unit interval.

Step 4 Repeat, . . .

If you've ever seen taffy being pulled, these processes (especially the second one) are essentially a one-dimensional version of taffy pulling. But making taffy is different in one fundamental way: you can't put two different parts of the taffy in exactly the same spot! It is not a two-to-one process. And that suggests a deep question: can a one-to-one function be chaotic? If the function $f : \mathbb{R} \to \mathbb{R}$ is continuous, then the answer is no. You should be able to prove this fairly easily.

But as we mentioned in section 11.6, the plane has more real estate and a function $F : \mathbb{R}^2 \to \mathbb{R}^2$ can twist and turn a planar region in a lot of different ways without mapping two points onto the same point. In this chapter, we introduce the prototypical example of a chaotic dynamical system that is one-to-one, onto, and continuous on an invariant set $\Lambda \subset \mathbb{R}^2$.

12.2 The Horseshoe Map

The first well-understood example of a chaotic dynamical having the properties alluded to above was developed by Steve Smale, a professor at the University of California, while he was on sabbatical in Rio de Janeiro in the early 1960s. Although the term "chaos" had not yet been defined in the way we use it now, Smale was essentially interested in constructing an example of a chaotic dynamical that was both invertible and hyperbolic (that is, persistent under small perturbations).

It is easiest and sufficient to describe the horseshoe map F in a step-by-step manner like we did for the doubling map and the logistic function above. The domain of the function is a rectangle with two half-disks attached to either end. I'll refer to this region as a "stadium" and denote it by S (see figure 12.1). Let $F : S \to S$ be defined as follows:

Step 1 Take S and compress it vertically to less than half its height.

Step 2 Take this compressed version of S and stretch it to twice its length.

Step 3 Fold this compressed and stretched version of S into a "horseshoe" shape and then place it back into S. This is $F(S)$.

Figure 12.1 illustrates the domain S and its image $F(S)$.

Explore

Let's begin to explore some of the basic properties of the function F.

Exploration 1. Explain in general terms why F is continuous.

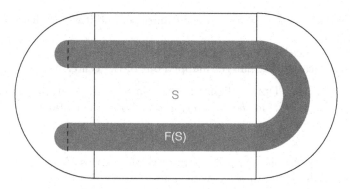

FIGURE 12.1
The Smale horseshoe map F. The domain of the function is the stadium-shaped region S. The image $F(S)$ is shown in gray.

Exploration 2. Explain in general terms why F is one-to-one.

Exploration 3. Explain in general terms why F is not onto S.

Exploration 4. Suppose that x is in the left semicircular region of S. Where is $F(x)$? What can you say about the orbit $\{F^n(x)\}$?

Exploration 5. Show that F has a unique attracting fixed point in the left semicircular region of S. What theorem did you use here?

Exploration 6. Suppose that x is in the right semicircular region of S. Where is $F(x)$? What can you say about the orbit $\{F^n(x)\}$?

Exploration 7. Are there points in the rectangular center region R whose orbits converge to the attracting fixed point that you just described?

Exploration 8. Sketch $F^2(S)$. It should have 4 thin horizontal strips through the center region R.

Exploration 9. Don't try and sketch $F^3(S)$ unless you are very patient. How many horizontal strips does it have through R? Generalize to $F^n(S)$.

These dynamics should remind of the logistic map with $a > 4$ that you studied in section 6.4. In that system, you showed the existence of a closed, invariant set $\Lambda \subset [0, 1]$ by first showing that there was a single open interval centered around $x = 1/2$ that left the interval in one iteration. Then you showed that there were 2 open intervals that mapped into this center interval and hence left in two iterations. And so on. You showed that what remained was a closed subset $\Lambda \subset [0,1]$ that is a Cantor set and if $x \in \Lambda$, then $f_a^n(x) \in \Lambda$ for all $n > 0$.

The same idea is happening here except instead of open intervals; there are open vertical strips that get mapped out of the center rectangle R.

Exploration 10. Draw a figure to explain why there are 3 vertical open strips U_ℓ, U_c, and U_r (for *left, center, and right*) in the rectangle R such that

$$F(U_i) \not\subset R$$

for $i \in \{\ell, c, r\}$. Explain why there are 2 vertical strips V_0 and V_1 in R such that $F(V_i) \subset R$. Label these on your figure.

Exploration 11. Now explain why there are 3 vertical strips in V_0 and 3 more in V_1 such that if x is in one of these 6 strips, then $F^2(x) \notin R$. How many strips are there whose iterates remain in R for 2 iterations?

Exploration 12. How many vertical strips are there in R such that if x is in one of these strips then $F^3(x) \in R$? In general, How many vertical strips are there in R such that if x is in one of these strips then $F^n(x) \in R$?

As in section 6.4, this construction continues for all n leaving a set

$$\Lambda_+ = \{x \in R \mid F^n(x) \in R \text{ for all } n \in \mathbb{N}\}.$$

Exploration 13. Describe the topological structure of the set Λ_+.

In exploration 2, you explained why the function F is one-to-one. And although F is not onto the entire domain (exploration 3), it is of course onto its image. This implies that F is invertible when its domain and range are appropriately restricted.

Exploration 14. Think about the original definition of the function F as described in the three-step process at the beginning of this section and illustrated in figure 12.1. Can you come up with a similar description for F^{-1}?

The invertibility of F is important. It tells us that the process used to define Λ_+ can be used with F^{-1} to define a similar set

$$\Lambda_- = \{x \in R \mid F^{-n}(x) \in R \text{ for all } n \in \mathbb{N}\}.$$

Exploration 15. Explain why Λ_- is a Cantor set of horizontal lines.

The set that we are interested in is the set of points that are in both Λ_+ and Λ_-. Define

$$\Lambda = \Lambda_+ \cap \Lambda_-.$$

Exploration 16. If $x \in \Lambda$, what can you say about both the forward and backward orbits of x?

12.3 More Symbolic Dynamics

Our ultimate goal here is to do exactly what we did with the logistic function and every other chaotic system that we studied earlier. We want to create a topological conjugacy between F restricted to the set Λ with a shift map σ defined on a symbol space Σ. In that way, one can prove that the horseshoe map F is, in fact, chaotic. We will omit the technical details needed to establish this conjugacy. If you are interested, you can read these details in *An Introduction to Chaotic Dynamical Systems* by Robert Devaney [2]:

12.3.1 Two-Sided Sequence Space

You probably noticed that we vaguely said "some shift map" and "some symbol space" and that was intentional. The invertibility of F allows us to use a variation on σ and Σ_2 that is more appropriate and therefore highlights some additional interesting dynamical properties. The key observation to defining the conjugacy, and hence the symbol space, is that the invariant set Λ lies entirely in the two vertical strips V_0 and V_1. Thus, just as we did in chapter 8, we want to associate with each $x \in \Lambda$ a sequence of 0s and 1s based on the itinerary of x. However, the invertibility of F implies that both the forward and backward orbits of x remain in Λ. In other words, the sequence should have both negatively and positively indexed entries. This leads to the following revised definition of the **doubly infinite sequence space** Σ_2:

$$\Sigma_2 = \left\{ \ldots s_{-2} s_{-1}.s_0 s_1 s_2 \ldots \mid s_j = 0 \text{ or } 1 \right\}. \tag{12.1}$$

Note that there is a dot (.) in these sequences between the negatively indexed entries and the non-negatively indexed entries. This dot, referred to as the **center** of the sequence, will be important in discussing dynamics on this space.

Explore

Exploration 17. Give examples of several different points in Σ_2.

This sequence space Σ_2 is again a metric space. However, the metric is slightly different than it was in our original sequence space. The metric d is defined by

$$d(s, t) = \sum_{k=-\infty}^{\infty} \frac{|s_k - t_k|}{2^{|k|}}. \tag{12.2}$$

You have probably not encountered an infinite series that runs from $-\infty$ to ∞ before. The easiest way to use this formula is to break it up into two different infinite series and then use the geometric series formula to evaluate each of them.

Exploration 18. Show that equation 12.2 can be rewritten as

$$d(s,t) = \sum_{k=1}^{\infty} \frac{|s_{-k} - t_{-k}|}{2^k} + \sum_{k=0}^{\infty} \frac{|s_k - t_k|}{2^k}.$$

Exploration 19. Let $s_1 = \ldots 00.11 \ldots, s_2 = \ldots 11.00 \ldots$ and $t = \ldots 00.00 \ldots$. Compute $d(s_1, t)$ and $d(s_2, t)$. Are they the same or different? Why?

Exploration 20. What is the maximal distance between any two points in Σ_2?

Prove

Proof 21. Prove that d, as defined in equation 12.2, is a metric. The properties of a metric are given in subsection 7.3.1.

Theorem 7.1 provided us with a natural way to see when two one-sided sequences s and t are near to each other. It stated that if s and t are two one-sided sequences such that $s_i = t_i$ for $i = 0, 1, \ldots, n$, then $d(s,t) \leq \frac{1}{2^n}$. This needs to be modified for two-sided sequences.

Proof 22. Prove the following theorem.

Theorem 12.1 *Let $s, t \in \Sigma_2$ and suppose that $s_i = t_i$ for $i = -n, \ldots, n$. Then*

$$d(s,t) \leq \frac{1}{2^{n-1}}.$$

Proof 23. Prove the following theorem.

Theorem 12.2 *If $d(s,t) < 1/2^n$ then $s_i = t_i$ for $i = -n, \ldots, n$.*

12.3.2   The Two-Sided Shift Map

As mentioned previously, our old shift map σ doesn't capture all of the properties of the horseshoe map. This is because that shift map is two-to-one and not one-to-one. For example,

$$\sigma(.0\overline{1}) = \sigma(.\overline{1}) = .\overline{1}.$$

But, in our new version of Σ_2, we can define a one-to-one version of the shift map $\sigma : \Sigma_2 \to \Sigma_2$ by

$$\sigma(\ldots s_{-2}s_{-1}.s_0 s_1 s_2 \ldots) = \ldots s_{-2}s_{-1}s_0.s_1 s_2 \ldots. \qquad (12.3)$$

Notice that the center (i.e., the "dot") has moved one unit to the right or equivalently, the sequence has *shifted* one unit to the left. Hence, the name *shift map*.

Explore

Exploration 24. What is the definition of σ^{-1}?

Exploration 25. What are the fixed points of σ? What are the period 2 points of σ? In general, what are the period n points of σ?

Because σ is invertible, the dynamical system defined by it is reversible; we can iterate both forward and backward. This leads to a refined definition of an orbit.

Definition 12.1 *The* **forward orbit** *of a point x_0 in a reversible dynamical system $x_{n+1} = f(x_n)$ is*

$$\mathcal{O}^+(x_0) = \{f^n(x_0) : n \geq 0\}.$$

The **backward orbit** *of a point x_0 is*

$$\mathcal{O}^-(x_0) = \{f^n(x_0) : n < 0\}.$$

The **orbit** *of x_0 is $\mathcal{O}(x_0) = \mathcal{O}^+(x_0) \cup \mathcal{O}^-(x_0)$.*

Exploration 26. Construct a point $s \in \Sigma_2$ having a dense forward orbit.

Exploration 27. Construct a point $s \in \Sigma_2$ having a dense backward orbit.

Exploration 28. Construct a point $s \in \Sigma_2$ having a dense forward orbit that does not have a dense backward orbit.

Exploration 29. Give an example to show that σ has sensitive dependence on initial conditions.

Exploration 30. Give an example to show that σ^{-1} has sensitive dependence on initial conditions.

Exploration 31. In section 11.6, we mentioned that homoclinic orbits are often an indicator of chaotic dynamics. Give an example of a point $s \in \Sigma_2$ that has a homoclinic orbit. More specifically, construct $s \neq \ldots 00.00 \ldots$ such that

$$\lim_{n \to \infty} \sigma^n(s) = \ldots 00.00 \ldots$$

and

$$\lim_{n \to -\infty} \sigma^n(s) = \ldots 00.00 \ldots .$$

Prove

Proof 32. Prove that σ as defined in equation 12.3 is one-to-one.

Proof 33. Prove that σ as defined in equation 12.3 is onto Σ_2.

Proof 34. Prove that periodic points of σ are dense in Σ_2.

Proof 35. Prove that σ has sensitive dependence on initial conditions.

Proof 36. Explain why σ is chaotic.

12.4   A Horseshoe in the Hénon Map

We motivated our investigation into the Smale horseshoe with a brief discussion of Hénon map given in equation 11.16. So let's finish this chapter by explicitly seeing this connection.

Exploration 37. Figure 12.2 shows a rectangle R and its image $H(R)$ by the Hénon map. Compare figure 12.2 with figure 12.1 which we used to define the Smale horseshoe map F. What are the similarities and differences?

Exploration 38. Figure 12.3 shows a rectangle R and its image $H^2(R)$ by the Hénon map. Compare figure 12.3 with the Smale horseshoe figure that you created in exploration 8. What are the similarities and differences?

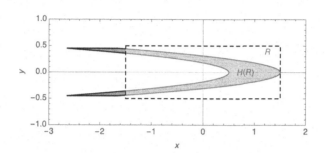

FIGURE 12.2
A region R (dashed box) and its image $H(R)$ via the Hénon map.

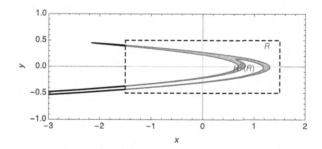

FIGURE 12.3
A region R (dashed box) and its image $H^2(R)$ via the Hénon map.

13

Generalized Symbolic Dynamics

13.1   Topology Foundations

In this section, we revisit the shift map from chapters 6, 7, and 12. We expand on those concepts to symbol spaces having more than two symbols. We will see how symbolic dynamics are not just a tool to study other dynamical systems, but are interesting dynamical systems in their own right. We also explore some of the tools use to study these systems. Students interested in a more complete overview of these topics should consult Ref. [6].

To begin, we introduce a few notions from topology that will be needed later. Recall from Chapter 7 the definition of a metric. Given a set M, a metric is a map

$$d : M \times M \to \mathbb{R},$$

which satisfies the following properties:

1. $d(s,t) = d(t,s)$.

2. $d(s,t) \geq 0$ for all s and t with equality if and only if $s = t$.

3. $d(s,t) \leq d(s,u) + d(u,t)$ for all s, u and t.

The ordered pair (M,d) is called a **metric space**.

We will be looking at shift dynamical systems similar to those that you already studied and will require an additional topological property on the symbol space known as compactness. The following definition is not the general topological definition of compactness, but in a metric space, it is equivalent.

Definition 13.1 *A metric space M is* **compact** *if every sequence in M has a convergent subsequence.*

We have used the term open set earlier in the text when we referred to open intervals or open sets in $\mathbb{R}^n$. The following definitions extends this concept to general metric spaces.

Definition 13.2 *Given a metric space (M,d), an* **open ball of radius ϵ, centered at a point** x *is the set*

$$B_\epsilon(x) = \{y \in M : d(x,y) < \epsilon\}.$$

A subset $U \subset M$ is **open** *if, for every $u \in U$, there exists some ϵ, such that $B_\epsilon(u) \subset U$.*

Additionally, both the metric space M and the empty set $\emptyset$ are defined to be open sets.

If we have an open set, we will also need a definition for a closed set.

Definition 13.3 *A subset $V \subset M$ is* **closed** *if its complement is open in M.*

In any metric space, there exists sets that are both open and closed (unimaginatively called "clopen"). Keep that in mind as you go through this chapter. Open sets and closed sets aren't mutually exclusive.

Sometimes, you start with an open set, and you want to make it closed.

Definition 13.4 *Given an open set U, its* **closure** *$\overline{U}$ is the smallest closed set containing U.*

The one theorem in this section is commonly proven in a course on real analysis or topology. If you haven't seen this proof before, it's worth giving it an attempt. The fact that this is the only proof assigned here should give you some insight into its use later in the chapter.

Prove

Proof 1. Let M be a compact metric space. A subset $N \subset M$ is compact if and only if N is closed.

Topology and Topological Spaces

The primary focus of topology is to determine when one mathematical object can be continuously deformed into another. For example, a cube can be continuously deformed into a sphere and so we say that these objects are "topologically equivalent." These mathematical objects are called **topological spaces**. Topological spaces consist of a set X and a collection of subsets $\mathcal{U}$ of X that are called the **open sets** of X. This collection of open sets must satisfy three axioms that specify how the sets of $\mathcal{U}$ relate to each other. All other properties of topological spaces derive from these axioms and other structures that might be inherent to X. For example, metric spaces possess an additional property that allows one to compute the distance between points in X and this gives additional structure to the space.

If you have not yet taken a topology course, we highly recommend taking one, especially if you are considering graduate school in mathematics.

13.2   Shift Dynamical Systems

In chapter 7, we defined the one-sided sequence space on two symbols (Σ_2) and the shift map $\sigma : \Sigma_2 \to \Sigma_2$. We then expanded this to a two-sided shift in chapter 12. Metric spaces such as these two are often referred to as **shift spaces**. Here, we generalize these ideas to sequences having more than two symbols and explore some of the uses of this generalization.

13.2.1   One-Sided Shift Spaces

We naturally adapt definition in 7.3.1 to involve n symbols. We still start our indexing at 0 so that

$$\Sigma_n = \{.s_0 s_1 s_2 \ldots \mid s_i \in \{0, 1, 2, \ldots, n-1\} \text{ for all } i\}. \qquad (13.1)$$

Following the same logic, we extend the metric defined in 7.3.1 to this space in the natural way. Define $d : \Sigma_n \times \Sigma_n \to \mathbb{R}$ by

$$d(s, t) = \sum_{k=0}^{\infty} \frac{|s_k - t_k|}{n^k}. \qquad (13.2)$$

Finally, we define the shift map $\sigma : \Sigma_n \to \Sigma_n$ by

$$\sigma(.s_0 s_1 s_2 \ldots) = .s_1 s_2 \ldots.$$

Explore

Exploration 2.   Prove that for all s, $t \in \Sigma_n$, $d(s,t) \le n$.

Exploration 3.   Describe the periodic points of the shift map σ in Σ_3. How many periodic points of period p are there?

Exploration 4.   Prove that the periodic points of the shift map σ are dense in Σ_3.

Exploration 5.   Prove that the shift map σ exhibits sensitive dependence on initial conditions Σ_3.

Exploration 6.   Construct a point $s \in \Sigma_3$ that has a dense orbit under iteration by σ.

Exploration 7.   Discuss how these questions do or do not generalize to Σ_n.

Exploration 8. Show that there is a one-to-one correspondence between the interval $[0, 1) \in \mathbb{R}$ and the one-sided shift on 10 symbols. What arithmetic operation in $\mathbb{R}$ would be represented by the shift map in this case?

13.2.2   Two-Sided Shift Spaces

We now move on to a two-sided shift, but this time we don't demand that our symbols be natural numbers.

Consider a finite set $\mathcal{A}$ of symbols, which we call the **alphabet**.

Definition 13.5 *The **full shift** on $\mathcal{A}$ is denoted*

$$\mathcal{A}^{\mathbb{Z}} = \{\ldots s_{-2}s_{-1}.s_0 s_1 s_2 \ldots : \ s_i \in \mathcal{A}\}. \tag{13.3}$$

The dot to the left of the 0-th term is the **origin**. In the case that the alphabet is the set $\{0, 1, \ldots, n-1\}$, the space is denoted the **full n-shift**.

A **word** or a **block** in $\mathcal{A}$ is any finite list of symbols chosen from $\mathcal{A}$. The number of symbols in a block u is its **length**, denoted $|u|$. Given $s \in \mathcal{A}^{\mathbb{Z}}$, we use $s_{[i,j]}$ to represent the block of s from the i-th to the j-th entry of s.

Again, you have encountered this concept previously in this book. Consider the point $s = \ldots 01.01 \ldots$ in the two-sided shift space Σ_2 of chapter 12. This sequence is constructed by recursively concatenating the word $u = 01$. Additionally, $|u| = 2$. Of course, this is not the only word in s. It might be helpful to list all of the words of length 2 and of length 3 that appear in s.

We will often want to consider shifts where certain words or blocks do not appear. To do that, we first let $\mathcal{F}$ be a set of words constructed from the alphabet $\mathcal{A}$ that we call the set of **forbidden blocks**. We then define $X_{\mathcal{F}} \subset \mathcal{A}^{\mathbb{Z}}$ to be the set of two-sided sequences $s \in \mathcal{A}^{\mathbb{Z}}$ which do not contain any of the words in $\mathcal{F}$. The set $X_{\mathcal{F}}$ is also a **shift space.**

The shift map from chapter 7 can be extended to shifts on spaces with forbidden blocks in the natural way, with $\sigma : \mathcal{A}^{\mathbb{Z}} \to \mathcal{A}^{\mathbb{Z}}$ given by

$$\sigma(\ldots s_{-2}s_{-1}.s_0 s_1 s_2 \ldots) = \ldots s_{-1}s_0.s_1 s_2 s_3 \ldots.$$

The action of the shift map is the same as before. What can be different is the underlying shift space. Because of this, it is common to use a subscript on the function name to indicate the underlying shift space.

When we use the term **shift dynamical system**, we mean a shift space X and an associated shift map σ_X. We write this as an ordered pair (X, σ_X).

It is sometimes the case that it is easier to describe a shift space by what is allowed, rather than what is not allowed (i.e., the set of forbidden blocks). Let X be a subset of a full-shift, and $B_n(X)$ the set of all n-blocks which appear in any element of X. The **language** of X is the set

$$\mathcal{B}(X) = \bigcup_{n=0}^{\infty} B_n(X).$$

Our next step is to construct a useful metric on these shift spaces. The generalization to arbitrary symbols, combined with a restricted language, makes our original metric in equation 13.2 problematic. So let's consider a different metric, $\rho : X_{\mathcal{F}} \times X_{\mathcal{F}} \to \mathbb{R}$ defined by

$$\rho(s,t) = \begin{cases} 2^{-k} & \text{if } s_{[-k,k]} \neq t_{[-k,k]} \text{ and } s_{[-i,i]} = t_{[-i,i]} \text{ for all } i < k\} \\ 0 & \text{if } s = t \end{cases}$$

(13.4)

In proof 28, you will prove that ρ satisfies the properties of a metric. For now, note that like all other metrics that we have used thus far, points are "close" if they agree near the origin (i.e., the decimal point).

It can be confusing when we try to write sequences of points in shift spaces since each point is also a sequence. Going forward, we'll be using a subscript to indicate the coordinate in a point and a superscript to indicate the index in a sequence of points. Thus, s_4 gives the 4th entry of the sequence s, s^4 gives the 4th element of the sequence $\{s^k\}$, and s_4^4 gives the fourth entry of the 4th element of the sequence $\{s^k\}$. This is consistent with how we denoted elements in a shift space in chapter 7, but inconsistent with the notation in chapter 2. We apparently can't make shift spaces without breaking some eggs, so we've made a choice of which egg to break.

For example, if we define the sequence $\{x^n\}$ in the one-sided shift on two symbols as the sequence of points of all terms 0, except for a 1 in coordinate n, we would have

$$x^0 = .10000\ldots$$
$$x^1 = .01000\ldots$$
$$x^2 = .00100\ldots$$

$$\vdots$$

Then, we would use the notation $x_0^2 = x_1^2 = 0$, and $x_2^2 = 1$. Remember, we start indexing the sequence at 0, starting to the right of the origin.

Explore

Exploration 9. Consider the full 2-shift. What is the alphabet? What are the forbidden blocks? Are you restricted in what symbols you use?

Exploration 10. Given an arbitrary element $s \in \mathcal{A}^{\mathbb{Z}}$, how would you calculate $|s_{[i,j]}|$?

Exploration 11. In the shift space $\{a, b, c\}^{\mathbb{Z}}$, find the distance ρ between the given points.

1. $\ldots abbac.abbac\ldots$ and $\ldots aabac.abbbc\ldots$
2. $\ldots abbac.abbac\ldots$ and $\ldots aabac.abcbb\ldots$
3. $\ldots abbac.abbac\ldots$ and $\ldots abbac.cbbac\ldots$

Exploration 12. Build a sequence in a shift space on two symbols following the construction below. Let

$$x_0 = \ldots 00.00\ldots$$
$$x_1 = \ldots 001.100\ldots$$
$$x_2 = \ldots 0011.1100\ldots$$

and so forth. To what does the sequence $\{x_n\}$ converge? Can you prove this using definition 2.2?

Exploration 13. Consider the shift space $\{0, 1, 2\}^{\mathbb{Z}}$, and the sequence

$$x_0 = \ldots 000.000\ldots$$
$$x_1 = \ldots 001.100\ldots$$
$$x_2 = \ldots 0012.2100\ldots$$
$$x_3 = \ldots 0120.0210\ldots$$
$$\vdots$$

List x_4 through x_6. Does this sequence converge? Why or why not?

Exploration 14. Is there a convergent subsequence of the sequence in exploration 13?

Exploration 15. Does the sequence in $\{0, 1, 2\}^{\mathbb{Z}}$ given below converge? Does it have a convergent subsequence?

$$x_0 = \ldots 1010.0101\ldots$$
$$x_1 = \ldots 2121.1212\ldots$$
$$x_2 = \ldots 0202.2020\ldots$$
$$x_3 = \ldots 1010.0101\ldots$$
$$\vdots$$

Given a shift dynamical $(\mathcal{A}^{\mathbb{Z}}, \sigma)$, and some subset $X \subseteq \mathcal{A}^{\mathbb{Z}}$, X is **shift invariant** if $\sigma(X) = X$. In other words, if x is any element in X, $\sigma(x) \in X$ as well.

Exploration 16. Are full shifts shift invariant? Are shift spaces shift invariant?

Exploration 17. Construct each of the following.

- A subset of the full-shift on 2 symbols which is shift invariant. This must contain more than 2 points.

- A subset of the full-shift on 3 symbols which is shift invariant. This must contain more than 3 points.

- A non-empty alphabet $\mathcal{A}$ such that every subset of shift space is shift invariant.

Exploration 18. Look at the shift space $\{0,1\}^{\mathbb{Z}}$. Show that the set

$$\{\ldots s_{-2}s_{-1}.s_0s_1 \ldots \mid s_{-2} = s_{-1} = s_0 = s_1 = 0 \text{ and } s_i \in \{0,1\} \text{ otherwise}\}$$

is both open and closed.

Exploration 19. Suppose you were to change the values on s_{-2} through s_1 in exploration 18. Pick a few different choices for 0 and 1 for each of these four terms around the origin. Show that the sets are still always open.

Exploration 20. For any number of the sets you described in exploration 19, show that their unions are also open.

Conjecture

Conjecture 21.

Suppose you have a shift space on n symbols, and a sequence $\{x_n\}$. Then sequence $\{x_n\}$ **must/must not** have a convergent subsequence.

Conjecture 22.

Suppose you have a shift space on n symbols, and a subset S of that space. Then S **must/must not** be shift invariant.

Conjecture 23. What does it mean for a set to be open in a shift space? Choose a shift we've already discussed, describe an open set, and conjecture a good definition for describing these sets in general.

To check the validity of your conjecture, verify that unions of open sets are open using your definition. Similarly, verify that finite intersections of closed sets are closed using your definition. (These are required properties of a topological space.)

Apply

In the course of proving Sarkovskii's Theorem, we noticed that when there was a period 3 orbit, there exists a pair of disjoint open intervals I_0 and I_1 such that $f(I_0) = I_1$ and $f(I_1) = I_0 \cup I_1$. Suppose that we assign a sequence of 0's and 1's to represent the itinerary of a point $x \in I_0 \cup I_1$. Call the set of sequences constructed in this manner X.

Application 24. Describe the set of forbidden words $\mathcal{F}$.

Application 25. List all words in $B_1(X)$? List all words in $B_2(x)$? List all words in $B_3(x)$?

Application 26. Describe the language $\mathcal{B}(X)$.

Apply

Application 27. In this application, you will prove that a shift space X is compact. The proof uses a variation of the Cantor diagonalization argument that is commonly used to prove that the real numbers are uncountable. You may wish to review that proof before working through the steps outlined below.

1. Assume you have a shift space X with an alphabet $\mathcal{A}$ and some sequence of points $\{x^n\}_{n=0}^{\infty}$ in X. It will be essential to remember that a shift space has a *finite* alphabet.

2. Now, consider the set $\{x_0^n\}$. This is a sequence of characters in $\mathcal{A}$, made from the 0-th coordinate of every point in the sequence $\{x^n\}_{n=0}^{\infty}$. Explain why there exists a $c_0 \in \mathcal{A}$ and an infinite set $N_0 \subset \mathbb{N}$ such that $x_0^k = c_0$ for all $k \in N_0$.

3. Next, show that there exists some infinite set $N_1 \subset N_0$, and some $c_{-1}, c_1 \in \mathcal{A}$, such that, for all $k \in N_1$, the block between index -1 and 1 satisfies
    $$x_{[-1,1]}^k = c_{-1}.c_0 c_1.$$

4. Continue in this manner to inductively construct the sets N_i and the associated blocks $x_{[-i,i]}^k$. What is the length around the origin on which all blocks agree for elements of $\{x^n\}$ with indices $n \in N_i$?

5. Let $c = \ldots c_{-2}c_{-1}.c_0c_1c_2\ldots$ constructed from the inductive steps described above. Explain why your inductive argument gives the existence of the point in $c \in X$.

6. Define the sequence
$$\{m_i\}_{i=0}^{\infty} \subset \mathbb{N}$$
with $m_0 = \min(N_0)$, $m_1 = \min(N_1)$, etc. Prove that the subsequence $\{x^{m_i}\}_{i=0}^{\infty}$ converges to c.

7. Explain why the existence of this convergent subsequence implies that X is compact.

Prove

Proof 28. Prove that the function ρ defined in equation 13.4 is a metric.

Proof 29. In chapter 7 you proved that the one-sided shift map on Σ_2 is continuous. Using the metric ρ defined in this chapter, prove that the shift map σ defined on $\mathcal{A}^{\mathbb{Z}}$ is continuous.

Proof 30. Let X be a shift space, u an element of the language $\mathcal{B}(X)$, and $k \in \mathbb{Z}$. A **cylinder set** in X is defined as the set
$$C_k^X(u) = \left\{ x \in X : x_{[k,k+|u|-1]} = u \right\}.$$

We interpret $C_k^X(u)$ as the set of all points which the block u occurs starting at index k. We say this cylinder set is **centered** at u and of **radius** k.

1. Prove that a cylinder set in a shift space is open.

2. Prove that a cylinder set in a shift space is closed.

Proof 31. Follow the outline below to show that a subset X of a $\mathcal{A}^{\mathbb{Z}}$ is a shift space if and only if it is shift invariant and compact.

1. Begin by assuming that X is a shift space. Use results from earlier in this section to show it is shift invariant and compact.

2. Now, assume that x is a shift invariant and compact subset of some full shift. What other property of this set do you have from theorem 1? What does this imply about the set $\mathcal{A}^{\mathbb{Z}} \setminus X$?

3. Show that, for any point $y \in \mathcal{A}^{\mathbb{Z}} \setminus X$, there must be a block u_y in y, of length $2k$, such that the cylinder set $C_{-k}^{\mathcal{A}^{\mathbb{Z}}}(u_y)$ is a subset of $\mathcal{A}^{\mathbb{Z}}$.

4. Let $\mathcal{F} = \{u_y : y \in \mathcal{A}^{\mathbb{Z}} \setminus X\}$. Prove that $X = X_{\mathcal{F}}$.

13.2.3 Shifts of Finite Type

Now, that we have discussed shift spaces, we will restrict ourselves to a specific kind of shift, called a shift of finite type. These show up regularly when we're finding shift spaces that are conjugate to dynamical systems.

Definition 13.6 *If the set $\mathcal{F}$ is finite, the shift space $X_{\mathcal{F}}$ is a* **shift of finite type (SFT)**.

Let's look at an example. Suppose we take a shift space on two symbols, 0 and 1 so that the alphabet is $\mathcal{A} = \{0,1\}$. We'll take our forbidden set of words to be $\mathcal{F} = \{000, 111\}$. Note that this rules out the case of more than three consecutive 0's or 1's. So the point $\ldots 01001.101101 \ldots$ is an element of $X_{\mathcal{F}}$, but the point $\ldots 01001.110101 \ldots$ would be forbidden.

We may need to add one additional restriction. When constructing points in a shift of finite type, we would like to determine what can occur next in a string, based on the past m elements.

Definition 13.7 *A shift of finite type X is an* **m-step shift** *if $X = X_{\mathcal{F}}$ where $\mathcal{F}$ contains only blocks of length $m + 1$.*

This allows us to consider only potential blocks of the same length as those in the set of forbidden blocks.

Explore

Exploration 32. What are the forbidden blocks in a full shift on n symbols?

Exploration 33. Is the empty set, $\emptyset$, a shift of finite type? Is it a shift space?

Exploration 34. In the example following definition 13.6, find m so that the shift space that is an m-step shift of finite type.

Exploration 35. Let $\mathcal{A} = \{a, b\}$ and $\mathcal{F} = \{bb\}$. Consider the 2-step shift of finite type $X_{\mathcal{F}}$.

1. Find all possible blocks of length 2. Since this is a 2-step shift, remember that all blocks of length greater than 2 must be made up of these blocks of length 2.

2. Find all of the blocks of length 3 in this space.

3. Find all of the blocks of length 4 in this space.

4. Could you reconstruct this space, changing the number of steps, where all blocks in $\mathcal{F}$ are of length 3? Of length 4? What do you think would be the advantage or disadvantage of doing so?

Exploration 36. Revisit chapter 8 and proofs 30 through 45 regarding the logistic map with $a > 4$. We'll build a an associated shift space.

1. What is the alphabet?
2. Is the shift one-side or two-sided?
3. What are the forbidden words?
4. How would you describe a general element of this shift space? Relate the shift map to the application of the logistic map with appropriate choice of a.

Exploration 37. Consider the shift space on $\mathcal{A} = \{0, 1\}$, where for any point in the space, each pair of 1's must be separated by an even number of zeros. So ... 100100.001 ... would be allowed, but ... 10010.001 ... would not. Is this a shift of finite type?

Conjecture

Recall the definition of a topological conjugacy from section 8.3 and consider this commutative diagram in figure 13.1. Assume that X is and Y is a shift space. $\mathcal{F}_1$ is the set of forbidden blocks in X, and $\mathcal{F}_2$ is the set of forbidden blocks in Y. We assume S is a conjugacy.

$$
\begin{array}{ccc}
X_{\mathcal{F}_1} & \xrightarrow{\;\sigma_X\;} & X_{\mathcal{F}_1} \\
\downarrow{\scriptstyle S} & & \downarrow{\scriptstyle S} \\
Y_{\mathcal{F}_2} & \xrightarrow{\;\sigma_Y\;} & Y_{\mathcal{F}_2}
\end{array}
$$

FIGURE 13.1
A commutative diagram between shift spaces $X_{\mathcal{F}_1}$ and $Y_{\mathcal{F}_2}$.

Conjecture 38. If a shift space is conjugate to a shift of finite type, must it be a shift of finite type?

Apply

Application 39. Begin with two intervals on the real line (which we will call *tiles*), labeled a and b. a has length $\frac{1+\sqrt{5}}{2}$ and b has length 1.

1. Place a copy of a to the left of the origin and a copy of b to the right of the origin.

2. Multiply the length of each tile by $\frac{1+\sqrt{5}}{2}$. You should think of this as stretching out the tiles, while keeping the origin between them. Explain why we can now relabel the expanded a tile as ab and the expanded b tile as a.

3. Repeat this process a few times. You should now be generating a two-sided sequence of the symbols a and b.

4. You have generated a **tiling** of the real line. Putting the origin between the a and b tiles is not special. It could be in the middle of a tile. Could it be between two of the same tile?

5. Look at the sequences you generated. Is there a set of forbidden blocks? Does this resemble a shift space you have already seen?

6. What might the relationship be between this geometric construction and a shift of finite type?

The Golden Ratio

The shift of finite type in exploration 35 is sometimes called the **golden ratio shift** or the **Fibonacci shift**.

The golden ratio is defined to be the ratio a/b that has the additional property

$$\frac{a}{b} = \frac{a+b}{a}.$$

This ratio is denoted φ and called the **golden ratio**. A little algebra shows that $\varphi = \frac{1+\sqrt{5}}{2}$.

The golden ratio shows up in a surprisingly wide variety of mathematical applications. For example, a regular pentagram inscribed in a regular pentagon exhibits the golden ratio. It has long been an important concept in art, architecture, and other disciplines. It is believed that the Parthenon in Athens was designed around the golden ratio.

One of the most interesting places that this remarkable constant arises is in the Fibonacci sequence where it can be shown that the ratio of consecutive Fibonacci numbers approaches φ. This is related to the fact that the golden ratio, when expressed as a continued fraction, has the form

$$\varphi = 1 + \cfrac{1}{1 + \cfrac{1}{1 + \cfrac{1}{1 + \cdots}}}.$$

For this reason, the shift space in exploration 35 is usually called the Fibonacci shift, and the tiling in application 39 is called the Fibonacci tiling.

13.3   Representing Shift Spaces with Graphs

It is possible to represent a shift of finite type visually with a finite directed graph. We discuss some of the basic notation here and illustrate their use with some applications of this technique.

We can think of a graph G as two sets, called vertices (V) and edges (E), where each edge has a direction associated with it so that you can "travel" from vertex to vertex within the graph by moving along connecting edges only in the indicated direction. If the sets V and E are relatively small, then a graph can be represented graphically with each vertex a circle and each edge a curve connecting the appropriate vertices. (Technically, we are describing what is known as a **directed graph** or **digraph**. However, in the context of dynamics, the directed edges are essential.)

In addition to the sets, we need two functions to keep track of our edges, which are called t and i. Both are functions from the set of edges to the set of vertices, and tell us where an edge ends (t for *terminal*) and where it starts (i for *initial*).

Another way of describing the connectivity of a graph is by using an **adjacency matrix**. An adjacency matrix A_G of a graph G is the matrix whose (m, n)th entry gives the number of edges which start at vertex m and end at vertex n.

Consider the following basic example. Let G be a graph with vertex set $V = \{0, 1\}$ and edge set $E = \{e_0, e_1, e_2\}$ and adjacency matrix

$$A_G = \begin{bmatrix} 0 & 1 \\ 1 & 1 \end{bmatrix}. \tag{13.5}$$

The graph G is shown in figure 13.2. In this example, the initial and terminal functions of edge e_0 are $i(e_0) = 0$ and $t(e_0) = 1$, respectively.

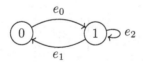

FIGURE 13.2
Digraph with adjacency matrix 13.5.

So, what does this have to do with dynamical systems and shift maps? We have a finite set of edges we're moving along and we can track which edges we use, moving forward or backward in time. Thus, tracking the sequence of edges traversed gives a shift space!

Definition 13.8 *Given a graph G, we define the* **edge shift** *as a set*

$$X_G = \{\ldots e_{-2} e_{-1}.e_0 e_1 e_2 \ldots \mid e_i \in E(G), t(e_i) = i(e_{i+1})\}.$$

The last condition checks that the edges in the sequence actually can follow each other in the directed graph.

Similarly, we can also track the sequence of vertices visited. This is known as a **vertex shift**.

Definition 13.9 *Let A be a $n \times n$ adjacency matrix with all entries either 0 or 1. Let G be its associated graph. The* **vertex shift** *is a shift space with alphabet $\{1, 2, \ldots, n\}$, denoted*

$$\widehat{X}_G = \{\ldots x_{-2} x_{-1}.x_0 x_1 x_2 \ldots \mid A_{x_i, x_{i+1}} = 1\}.$$

Explore

Exploration 40. Draw the graph for each of the following adjacency matrices. Describe the edge shift for each graph. Describe the vertex shift for each graph.

1.
$$\begin{bmatrix} 1 & 1 \\ 1 & 1 \end{bmatrix}$$

2.
$$\begin{bmatrix} 0 & 1 \\ 2 & 0 \end{bmatrix}$$

3.
$$\begin{bmatrix} 1 & 2 & 1 \\ 1 & 0 & 1 \\ 1 & 0 & 0 \end{bmatrix}$$

Exploration 41. How is the adjacency matrix given in equation 13.5 and its associated graph in figure 13.2 related to a period 3 orbit of a continuous dynamical system on $\mathbb{R}$?

Exploration 42. For each edge shift in exploration 40, can you find either a language or a set of forbidden blocks? What about for the vertex shift?

Exploration 43. Construct an example of a graph that does not have an edge shift. Find its adjacency matrix.

Exploration 44. For the graph associated with equation 13.5, find the eigenvalues of the adjacency matrix. What do you suppose is the relationship between these values, the shift space in exploration 35 and the tiling in application 39?

Exploration 45. We claim there is no edge shift conjugate to the shift of finite type in exploration 35. If each of the symbols in exploration 35 were an edge, how many vertices would there be? Could you find a word in the edge shift that is forbidden in the shift space?

Conjecture

Conjecture 46.

An edge shift with a single vertex and n edges is conjugate to ⎯⎯⎯⎯⎯⎯.

Conjecture 47. Take a graph from an earlier exercise, and it's adjacency matrix A. We know what information about the graph is available from A. Calculate A^2, A^3, and A^4. What information about the graph is contained in A^k? In particular, what do the diagonal entries compute?

Apply

Application 48. In exploration 41, you made the connection between period 3 orbits and the adjacency matrix of equation 13.5. How can you use conjecture 47 to count the number of period k orbits?

Application 49. The *trace* of a matrix M is the sum of the diagonal entries of M. For the adjacency matrix of equation 13.5, compute the trace of A^k for $k = 1, 2, \ldots, 5$. What is the pattern of these numbers? What does that tell you about periodic orbits?

13.3.1   Higher Edge Graphs

We saw in exploration 45 that an edge shift, which shares all of the same symbols as a shift of finite type may not account for the entire language or all of the forbidden words. The higher block code and higher edge graph is a tool developed to handle this limitation.

We denote a **higher block code** for a shift X as $X^{[N]}$. By this, we mean a shift constructed on a new alphabet, whose elements are the blocks of length N in the original shift space X.

Definition 13.10 *Given a graph G and $N \geq 1$, the N-th* **higher edge graph** *$G^{[N]}$ is a graph whose vertex set is the collection of paths of length $N - 1$ in G. Let $e_1^1 e_2^1 \ldots e_{N-1}^1$ and $e_1^2 e_2^2 \ldots e_{N-1}^2$ be two vertices in $G^{[N]}$. We define exactly one edge in $G^{[N]}$ whenever $e_2^1 \ldots e_{N-1}^1 = e_1^2 e_2^2 \ldots e_{N-2}^2$. In the case $N = 2$, this condition is replaced with $t(e_1^1) = i(e_1^2)$.*

Note that in a higher edge graph, there is at most one edge between any pair of vertices. We can use this when creating a shift directly from an adjacency matrix.

Let's look at an example to motivate this definition. Start with the graph G in figure 13.3. This should look familiar.

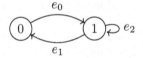

FIGURE 13.3

It has vertex set 0 and 1. The first higher edge graph $G^{[1]}$ has a vertex set made up of paths of length 0. We'll say a path of length 0 is just a vertex so that by definition $G^{[1]} = G$.

Now, construct $G^{[2]}$. The paths of length 1 in G define the new vertex set. In other words,

$$V\left(G^{[2]}\right) = \{e_0, e_1, e_2\}.$$

Definition 13.10 tells us how to construct the edges between these vertices. From the graph of G, we see that after traversing e_0, we could either go on edge e_1 or edge e_2. In other words,

$$t(e_0) = i(e_1) = i(e_2).$$

So, we need an edge from *node* e_0 to *node* e_1 and a second edge from *node* e_0 to *node* e_2. We label these edges $e_0 e_1$, and $e_0 e_2$. Similarly, we need edges $e_1 e_0$, $e_2 e_1$, and $e_2 e_2$. This gives the graph $G^{[2]}$ as shown in figure 13.4.

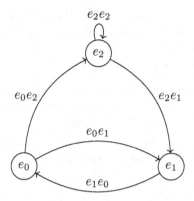

FIGURE 13.4
Second higher edge graph of the graph G in figure 13.2.

Explore

Exploration 50. Using the example from figures 13.3 and 13.4, construct $G^{[3]}$. To get you started, the nodes of this graph are the edges of $G^{[2]}$ and so there are 5 of them.

Exploration 51. Write a few elements from the shift spaces associated with G, $G^{[2]}$, and $G^{[3]}$. Does taking a higher edge graph cause any forbidden words to appear in elements of the shift space?

Exploration 52. For each graph in exploration 40, find the alphabet of the associated edge shift. You can use any labeling you like. Find the alphabets for the 2nd and 3rd higher block codes for each shift you've described.

Exploration 53. For each graph in exploration 40, find the adjacency matrix of the 2nd higher edge graph.

Exploration 54. We saw in exploration 45 that the shift of finite type in exploration 35 was not conjugate to the edge shift we'd looked at earlier.

1. Call the shift in exploration 35 X. Write out the elements of the alphabet for the higher block code $X^{[2]}$.

2. Show that, with careful relabeling, this is conjugate to an edge shift on the graph G from the start of this section.

Exploration 55. Draw a directed graph using at least 3 vertices and at least 6 edges. No vertex should be isolated and you should have at least once edge leaving each vertex. Self-loops are allowed.

In a graph, a **cycle** is a path that starts and ends at the same vertex.

1. Write the adjacency matrix A for the graph you've drawn. Calculate the trace of A (the sum of the diagonal entries) and count the number of self-loops you've included.

2. Multiply the matrix by itself to get A^2. For each vertex, count the number cycles using exactly two edges. Compute the trace of A^2.

3. Try this with a few higher powers on the matrix. For each A^k, calculate by hand the number of cycles of length k for each vertex, then calculate the trace of A^k.

4. How does this relate to periodic orbits?

Conjecture

Conjecture 56.

For $N \geq 2$, only the numbers _____ will appear in the adjacency matrix for $G^{[N]}$.

Conjecture 57.

The trace of the k-th power of the adjacency matrix gives you the number of _____ in the associated edge shift.

Apply

Application 58. Draw the directed graph G for the adjacency matrix

$$\begin{bmatrix} 2 & 3 \\ 1 & 1 \end{bmatrix}.$$

Label the vertex on the left v and the vertex on the right w. Label the two self-loops on v as a and b and the three edges from v to w as c, d, and e. Label the edge from w to v as f and the self-loop on w as g.

1. Introduce a probability function $P(G) \to [0,1]$. Let $P(v) = 1/3$ and $P(w) = 2/3$.

2. Repeatedly use the random number generator on the website to choose a value between 1 and 3. This process generates an element of the shift space associated with this graph. Think of this as moving through the graph. If you get a 1, you're at v, if you get a 2 or 3, you're at w.

3. Generate a few elements of the vertex shift, using these probabilities to choose which vertex is next in the sequence. Are there any forbidden words?

4. We next want to assign probabilities to each edge. We first introduce the idea of conditional probability. Let $P(E|V)$ be the probability of taking edge E given that you are at vertex V.

   (a) Which conditional probabilities need to be zero?

   (b) Choose probabilities for each edge such that

   $$\sum_{E \in E(G)} P(E|V) = 1 \text{ for all } V \in V(G).$$

   (c) Use the random number generator on the website to construct edge sequences. Are there any forbidden words? Do you see any relationship between the vertex sequences and the edge sequences?

5. Define the probability of an sequence of n edges to be

   $$P(E_1 E_2 \ldots E_n) = P(i(E_1))P(E_1|i(E_1))P(E_2|i(E_2)) \cdots P(E_n|i(E_n)).$$

6. Show that the set of possible sequences are exactly the blocks in the language for the edge shift X_G.

What you have constructed is an example of a **Markov Chain**.

Prove

Proof 59. Let G be a graph. Recalling our notation, X_G is the associated edge shift and $(X_G)^{[N]}$ is the N-th higher block code of that shift. $G^{[N]}$ is the N-th higher edge graph and $X_{G^{[N]}}$ is its associated edge shift. Prove that

$$(X_G)^{[N]} = X_{G^{[N]}}.$$

Proof 60. Prove that, up to renaming of symbols, the 1-step shifts of finite type are vertex shifts.

Proof 61.   Prove that, up to a renaming of symbols, every edge shift is a vertex shift (of a different graph).

Proof 62.   Prove that if X is an M-step shift of finite type, there is a graph G such that $X^{[M]} = \widehat{X}_G$ and $X^{[M+1]} = X_G$. That is, the M-th higher block code is the same as the vertex shift of G and the $M + 1$-st higher block code is the same as the edge shift of G.

Proof 63.   Prove that the i, j entry of the k-th power of an adjacency matrix gives the number of paths involving k edges when moving from vertex i to vertex j.

Proof 64.   Prove that the trace of the k-th power of the adjacency matrix gives you the number of points of period k in the associated edge shift.

13.4   Markov Partitions

In this section, we will assume we have a discrete dynamical system on a metric space X given by some function $f : X \to X$. We will also assume that the map f is invertible so that the notation $f^n(x)$ is defined for any $x \in X$ and any integer n. We write (X, f) to denote such a dynamical system.

The idea in this section is to divide X into a finite number of parts and keep track of what part a point in X lands in after n iterations of the dynamical system. We are tracking the orbit of a point, as in definition 12.1, and looking at where points land, as we did in sections 8.2 and 12.2.

Definition 13.11 *A* **topological partition** $\mathcal{P}$ *on a metric space X is a finite collection of disjoint open sets* $\mathcal{P} = \{P_1, P_2, \ldots, P_n\}$ *, whose closures cover X. That is, $X = \overline{P_1} \cup \cdots \cup \overline{P_N}$.*

Once we have such a partition, we can use it to build a symbolic representation of a dynamical system.

Definition 13.12 *Suppose that we have a dynamical system (X, f), a topological partition $\mathcal{P}$ on X, and indexing set $\{1, 2, \ldots, n\}$ for the open sets in the partition. A word $a_1 a_2 a_3 \ldots a_r$ is* **allowed for** $\mathcal{P}$ *if*

$$\bigcap_{j=1}^{r} f^{-1}(P_j) \neq \emptyset.$$

Definition 13.13 *Let $\mathcal{L}_{\mathcal{P},f}$ be the set of all allowed words for partition $\mathcal{P}$ on (X, f). The shift space $X_{\mathcal{P},f}$ with language $\mathcal{L}_{\mathcal{P},f}$ is the* **symbolic dynamical system corresponding to** $\mathcal{P}$.

There is an issue in definition 13.12 associated with the use of f^{-1}. In particular, the inverse images of single points may not be unique. However, we will restrict our work here only to invertible functions to avoid this complication.

For a given invertible dynamical system (X, f) and a topological partition $\mathcal{P}$ for each $x \in X_{\mathcal{P},f}$, define

$$D_n(x) = \bigcap_{j=-n}^{n} f^{-1}(P_{x_j}) \subseteq X.$$

Remember that $X_{\mathcal{P},f}$ is a symbolic dynamical system, so the coordinate x_j indicates which set in the topological partition the point x ended up after j iterations of f.

Definition 13.14 *For an invertible dynamical system (X, f), the topological partition $\mathcal{P}$ gives a* **symbolic representation** *of (X, f) if for every $x \in X_{\mathcal{P},f}$ the set*

$$\bigcap_{n=0}^{\infty} \overline{D_n(x)}$$

is a single point. If, in addition, $X_{\mathcal{P},f}$ is a shift of finite type, $\mathcal{P}$ is a **Markov partition***.*

Explore

Exploration 65. Let M be the unit circle. We think of this as just the interval $[0, 1)$, but where any function acting on a point in M has its result taken mod 1 as we did with the doubling map D.

Define the dynamical system $\phi : M \to M$ by $\phi(x) = 10x \pmod 1$.

- Show that subdividing M into the intervals $(\frac{i}{10}, \frac{i+1}{10})$ for $i \in 0, 1, \dots, 9$ is a topological partition.

- For a given point in M, show that tracking which set in the partition a point lands under this map gives an element of a one-sided shift. Where do you encounter an issue?

- Is this a Markov partition?

- Relate this to the shift you described in exploration 8.

Exploration 66. Given a symbolic representation of a dynamical system (X, f), define $\pi : X_{\mathcal{P},f} \to X$ to be the map which sends a point in the symbolic representation to the unique point in $\bigcap_{n=0}^{\infty} \overline{D_n(x)}$. Prove that the following diagram commutes.

$$X_{P,f} \xrightarrow{\sigma} X_{P,f}$$

$$\downarrow{\pi} \qquad \downarrow{\pi}$$

$$X \xrightarrow{f} X$$

Apply

Application 67. If you have not done so, we recommend you review the materials in section 11.1 prior to working out this application.

Define the torus $\mathbb{T}^2 = \mathbb{R}^2/\mathbb{Z}^2$. You can think of this as the subset of the plane $[0,1) \times [0,1)$ where any linear transformation applied to points in the set is resolved by taking each coordinate modulo 1.

1. Define a linear transformation on $\mathbb{R}^2$ with the matrix

$$A = \begin{bmatrix} 1 & 1 \\ 1 & 0 \end{bmatrix}.$$

   Compute the eigenvalues and eigenvectors of this matrix.

2. Let

$$\phi(x,y) = A\begin{pmatrix} x \\ y \end{pmatrix},$$

   with addition modulo 1. Let $\pi : \mathbb{R}^2 \to \mathbb{T}^2$ be given by

$$\pi(x,y) = (x \pmod 1, y \pmod 1).$$

   Prove that the diagram below commutes.

$$\mathbb{R}^2 \xrightarrow{A} \mathbb{R}^2$$

$$\downarrow{\pi} \qquad \downarrow{\pi}$$

$$\mathbb{T}^2 \xrightarrow{\phi} \mathbb{T}^2$$

3. Let L_1 be the span of the vector

$$\mathbf{v}_1 = \begin{pmatrix} 1 \\ -\frac{1-\sqrt{5}}{2} \end{pmatrix}$$

   and L_2, the span of the vector

$$\mathbf{v}_2 = \begin{pmatrix} 1 \\ -\frac{1+\sqrt{5}}{2} \end{pmatrix}.$$

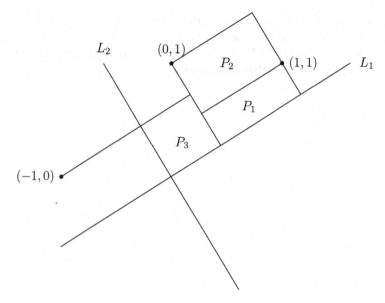

4. Begin with the three rectangles P_1, P_2, P_3 in the figure below. Describe how they were drawn.

5. Let $P_i' = \pi(P_i)$. This gives us the following figure. Explain how this figure was drawn and prove that this is a Markov partition for dynamical system $(\mathbb{T}^2, \phi)$.

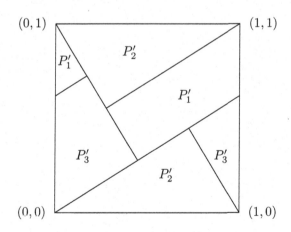

Application 68. Perform the same process as above, using the matrix

$$\begin{bmatrix} 2 & 1 \\ 1 & 0 \end{bmatrix}.$$

Construct the Markov partition and the symbolic representation of the map on the torus. Show that this map is chaotic.

This is more commonly known as **Arnold's cat map** and is an example of a **hyperbolic toral automorphism**.

Bibliography

[1] Robert Devaney. *A first course in chaotic dynamical systems: Theory and Experiment*. Perseus Books, 1992.

[2] Robert Devaney. *An introduction to chaotic dynamical systems*. CRC Press, 2018.

[3] James Gleick. *Chaos: Making a new science*. Open Road Media, 2011.

[4] Victor Guillemin and Alan Pollack. *Differential topology*, volume 370. American Mathematical Soc., 2010.

[5] TY Li and JA Yorke. Period 3 implies chaos. *Am. Math. Mon.*, 82:985–992, 1975.

[6] Douglas Lind and Brian Marcus. *An introduction to symbolic dynamics and coding*. Cambridge University Press, 1995.

[7] EN Lorenz. Deterministic nonperiodic flow. *J. Atmos. Sci.*, 20:130–141, 1963.

[8] AN Sharkovskiĭ. Coexistence of cycles of a continuous map of the line into itself. *Int. J. Bifurcat. Chaos*, 5(05):1263–1273, 1995.

Index

Printed in the United States
By Bookmasters